AF346303

From Dichotomy to Harmony: The Evolving View of the Immune System's Role

Narayan

Copyright © 2024 by Narayan

All rights reserved. No part of this book may be
reproduced in any manner
whatso-ever without written permission except in the
case of brief quotations
embodied in critical articles and reviews.

First Printing, 2024

Table of contents

Chapter 1: Introduction

1.1. General introduction

The immune system is a defense system tasked with protecting the body against environmental pathogens and other threats to its integrity. To maintain homeostasis, the immune system relies on molecular and cellular actors to integrate and respond to environmental signals. While immunity is crucial for surviving in an infected world, excessive immune responses are harmful and have to be controlled.

Traditionally, the immune system was thought to protect 'self' from 'non-self'[1]. This dichotomy has been replaced by the concept of 'meta-organism', in which the host and associated microbial communities form an interdependent symbiotic unit[2,3]. Rather than protecting 'self' from 'non-self', the immune system maintains homeostasis by favoring cooperators to the system and eliminating defectors, which englobe both external threats, like pathogens, and internal threats, like tumor cells.

1.2. Basic concepts in immunity

The immune system is a multilayered defense system, which includes innate and adaptive immune responses. Innate and adaptive immune responses, despite relying on different molecular and cellular actors, and differing in kinetics, specificity, and recognition systems, collaborate to maintain immune homeostasis[4]. The interplay between innate and adaptive responses requires dynamic communication and relies on contact-dependent and contact-independent interactions[5–7].

1.2.1. Innate immunity

To penetrate the organism, pathogens have to cross several barriers (**Figures 1, 2**). Skin and mucosal surfaces are lined with tightly joined epithelial cells, which form a physical barrier to pathogen penetration[8]. Physical barriers also include cilia, which are found on the respiratory epithelium and contribute to pathogen clearance in the lungs[9,10]. Additional chemical and biological barriers also protect mucosal surfaces, which are in contact with the environment, and therefore constitute potential entry points for pathogens. Chemical barriers rely on low pH, as well as enzyme, mucus, and antimicrobial peptide production.

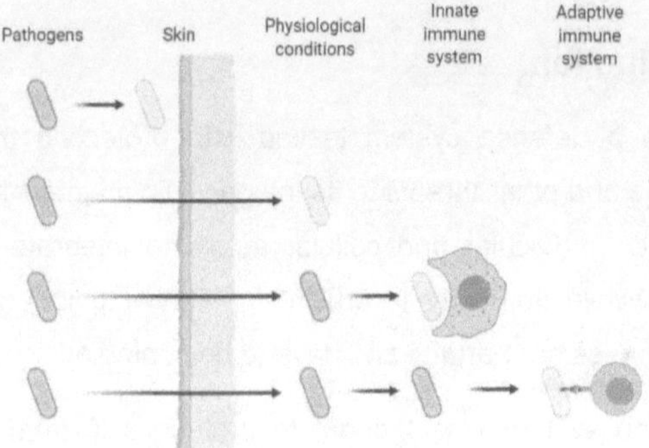

Figure 1. **The body has several lines of defense against pathogens.**
To penetrate the organism, pathogens first have to cross physical barriers, like the skin and other epithelial surfaces. Pathogen penetration is subsequently hindered by chemical and biological barriers, like low pH, enzyme production, or the microbiota. Pathogens that manage to cross these different barriers are detected by the innate immune system, which is able to eliminate a wide range of pathogens. More resistant pathogens are handled by the adaptive immune system. All these layers collaborate to ensure maximal protection against pathogens.

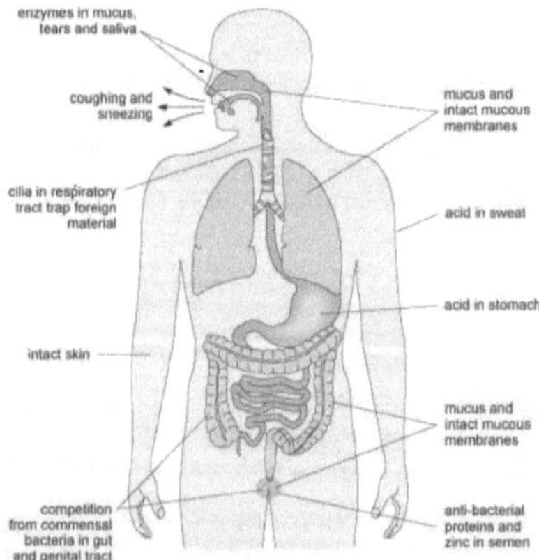

Figure 2. **Physical and chemical barriers against infection.**
To penetrate the organism and reach the immune system, pathogens have to cross several physical and chemical barriers, some of which are mentioned in this figure.

This is notably the case in the stomach, with gastric acid and gastric enzymes, and in the intestinal environment, where the mucus layer and antimicrobial peptides help contain the microbiota[11,12]. The microbial species forming the microbiota constitute a biological barrier and prevent pathogen invasion by competing for niche occupation[13]. Once these barriers have been crossed, pathogens come into contact with innate immune defenses (**Figure 1**).

Innate immunity is an evolutionarily ancient form of defense found in all animals and plants[14,15]. It acts quickly and allows near-immediate protection from a wide range of pathogens[16]. Innate immunity comprises multiple cell types: dendritic cells (DC), macrophages, mast cells, neutrophils, eosinophils, basophils, and natural killer (NK) cells (**Figure 3**). Innate immune cells recognize pathogen families, rather than specific pathogens, and are able to alert other components of the immune system as soon as an infection occurs. They express Pattern Recognition Receptors (PRRs), which are germline-encoded receptors that detect Pathogen-Associated Molecular Patterns (PAMPs)[17]. PAMPs are conserved molecular structures expressed by pathogens of a given type. Among PRRs, Toll-like receptors (TLRs) form a family of membrane-bound receptors that recognize a variety of ligands[18] (**Figure 4**). For example, TLR4 recognizes lipopolysaccharide (LPS) in Gram-negative bacteria, TLR2 recognizes lipoteichoic acid found on the cell wall of Gram-positive bacteria, and TLR3 binds to viral double-stranded RNA. Another family of membrane-bound receptors, the C-type lectin receptors (CLRs), are instrumental in recognizing fungi[19]. In the cytoplasm, PRRs include NOD-like receptors (NLRs)[20], as well as RIG-I-like receptors (RLRs), which are RNA helicases that recognize viral RNA[21]. PRRs can also sense cellular damage, by detecting Danger-Associated Molecular Patterns (DAMPs). DAMPs are normal cellular components, such as ATP, DNA, RNA, and histones that are released upon cell lysis, occurring after infection or trauma[22]. Innate immune cells express a vast array of PRRs. Signals originating from different PRRs are integrated by innate immune cells and lead to the development of different immune responses[23].

Successfully detected pathogens will consequently be eliminated by innate immune cells. Many of these cells rely on phagocytosis to eliminate the pathogens that they encounter.

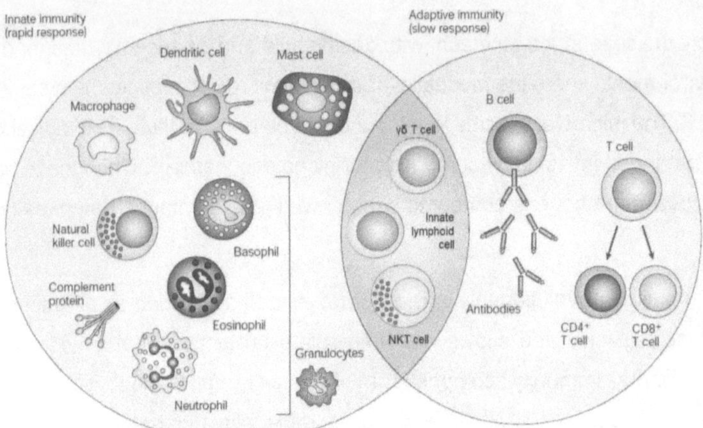

Figure 3. **Innate and adaptive immune cell types.**
Innate immunity acts as a first line of defense against infection. It comprises several cell types: granulocytes (basophils, eosinophils, and neutrophils), mast cells, macrophages, dendritic cells, and natural killer cells. The complement system also belongs to the innate immune system. The adaptive immune response is slower, but more specific and leads to the development of immune memory. It consists of antibodies, B cells, and CD4⁺ and CD8⁺ T lymphocytes. Natural killer T cells, innate lymphoid cells, and γδ T cells are lymphocytes that exhibit an intermediate phenotype between innate and adaptive immunity.
Adapted from: Dranoff, Glenn. "Cytokines in Cancer Pathogenesis and Cancer Therapy." *Nature Reviews Cancer* 4, no. 1 (January 2004): 11–22. https://doi.org/10.1038/nrc1252.

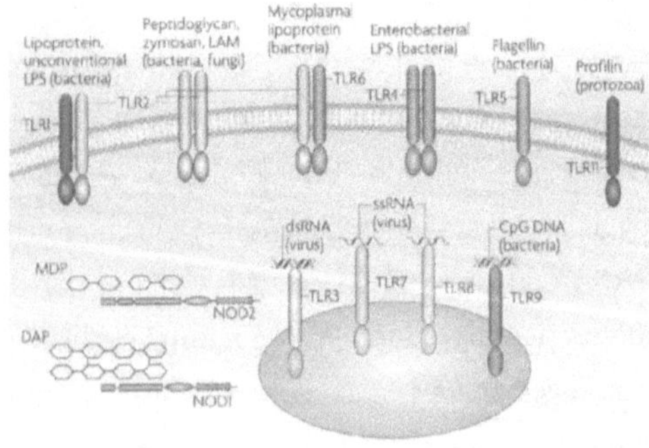

Figure 4. **Pattern-recognition receptors.**
Pattern Recognition Receptors (PRRs) are germline-encoded receptors that detect conserved molecular structures expressed by pathogens of a given type (PAMPs). The figure focuses on the better-known pattern-recognition receptors — Toll-like receptors (TLRs) and NODs. PRRs can be membrane-bound or cytoplasmic. DAP, diaminopimelic acid; ds, double-stranded; MDP, muramyl dipeptide; LPS, lipopolysaccharide; LAM, lipoarabinomannan; ss, single-stranded.
From: Kaufmann, Stefan H. E. "The Contribution of Immunology to the Rational Design of Novel Antibacterial Vaccines." *Nature Reviews Microbiology* 5, no. 7 (July 2007): 491–504.

Phagocytosis is the process by which hematopoietic cells engulf and subsequently lyse extracellular elements, such as pathogens[24–26] (**Figure 5**). Pathogens entering the body are first recognized by endothelial cells or resident macrophages, which initiate the inflammatory response and attract phagocytes, such as neutrophils, from the bloodstream[27–29]. Certain immune cells, like mast cells and granulocytes (eosinophils, basophils, and neutrophils) eliminate pathogens by releasing cytosolic granules, which contain antimicrobial agents, enzymes, or reactive oxygen species[30,31]. NK cells are the main innate mediator of cytotoxicity[32]. They induce cell death by releasing lytic granules or by expressing death receptor ligands, which provoke cell apoptosis when they bind to the appropriate receptor at the surface of pathogen cells. Innate immunity also relies on the complement system[33,34], blood-circulating protease-activated proteins that can potentiate phagocytic activity, promote inflammation, and directly attack pathogen cell membranes.

Innate immunity is often sufficient to protect the body against pathogens, but it has its set of shortcomings. The recognition ability of innate immune cells is limited. Fast mutation rates and intracellular replication make pathogen detection and elimination even more challenging. Moreover, innate immune responses are stereotyped and lack memory capacity[4].

1.2.2. Adaptive immunity

Jawed vertebrates have evolved an additional immune system, with improved pathogen recognition: the adaptive immune system. Adaptive immune cells are characterized by the expression of antigen-specific receptors, which have the potential to recognize all known molecular structures[35]. This gives them the ability to specifically recognize pathogens and adapt to their fast evolution rate. The cells of the adaptive immune system are T and B lymphocytes (**Figure 3**). They express T cell receptors (TCR) and B cell receptors (BCR), respectively (**Figure 6**). BCR and TCR are generated from a limited set of genes, called V(D)J genes, which are randomly assembled by a process of somatic recombination[36,37], endowing them with a virtually unlimited diversity (> 10^{15} antigens)[38].

The adaptive immune system is shaped by the innate immune system, through the action of antigen presenting cells (APC), such as dendritic cells[39–41]. Immature DCs reside in

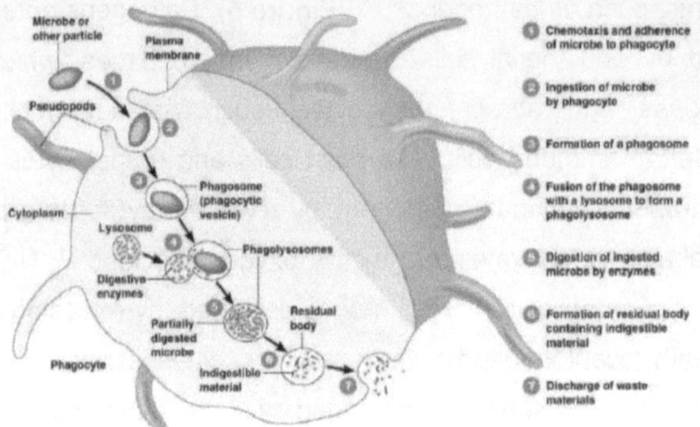

Figure 5. **Phagocytosis.**
Phagocytosis is the process by which hematopoietic cells engulf and subsequently lyse extracellular elements, such as pathogens, as described in this figure.
From: Tortora, Gerard J., Berdell R. Funke, and Christine L. Case. Microbiology: An Introduction. Pearson Benjamin Cummings, 2007.

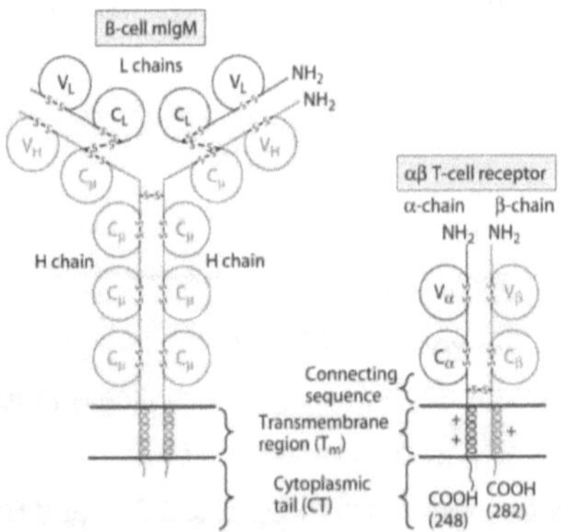

Figure 6. **B cell receptor (BCR) and T cell receptor (TCR) structure.**
The B cell receptor (BCR) and the T cell receptor (TCR) are transmembrane proteins present on the surface of B cells and T cells, respectively, which control the activation of these cells. BCRs contain heavy (H) and light (L) chains, and TCRs contain an α and a β chain. These chains are composed of variable (V) and constant (C) regions.
From: Kindt, Thomas J., Richard A. Goldsby, Barbara A. Osborne, and Janis Kuby. Kuby Immunology. W. H. Freeman, 2007.

tissues and constantly sample the surrounding environment for pathogens. After phagocytosing a pathogen, DCs degrade their proteins into peptides. DC maturation is subsequently triggered by the detection of PAMPs. Mature DCs present processed antigens on their surface via MHC molecules. MHC molecules are transmembrane presenting molecules made up of two chains that fold together and form a long cleft in which the peptide nests[42,43]. Mature DCs migrate to the secondary lymphoid organs, where they present antigens to T cells, and can thereby activate adaptive immunity.

While T cells get activated by recognizing their cognate antigen on the surface of a DC, B cells are able to recognize unprocessed antigens. Cell activation leads to clonal expansion, thus generating a large pool of specific T and B cell clones, which all recognize the same antigen[44–48] (**Figure 7**). B cells are responsible for the induction of humoral responses. They act remotely by producing antibodies that target pathogenic antigens and can travel via the circulating system. Antibodies can neutralize toxins, activate the complement system, or promote phagocytosis and cytotoxicity[49–51]. After activation, T cells migrate to the infection site and exert their effector functions. T cells are responsible for cellular responses and can be classified into two subsets, based on their expression of coreceptors CD8 and CD4. CD8+ cytotoxic T lymphocytes (CTLs) are able to directly kill infected or damaged cells by releasing cytotoxins, like perforin and granzyme[52,53]. Perforin allows the intracellular delivery of granzyme and other proteases into the cytoplasm of the target cell, leading to cell lysis[30]. CTLs also secrete soluble mediators that interfere with viral replication. CD4+ helper T lymphocytes orchestrate immune responses by enhancing the function of other immune cell types through a combination of contact-dependent and contact-independent interactions, like secreting cytokines and other soluble mediators[54,55].

The adaptive immune response is characterized by its ability to develop immune memory. First time exposure to a pathogen leads to the development of a primary immune response. This response develops slowly and produces mostly short-lived cells, which help resolve the infection. Long-lived memory cells also develop during primary immune responses[4,45] (**Figure 7**). They are able to initiate a greater and faster immune response after a second exposure to the same pathogen, establishing a state of memory, which

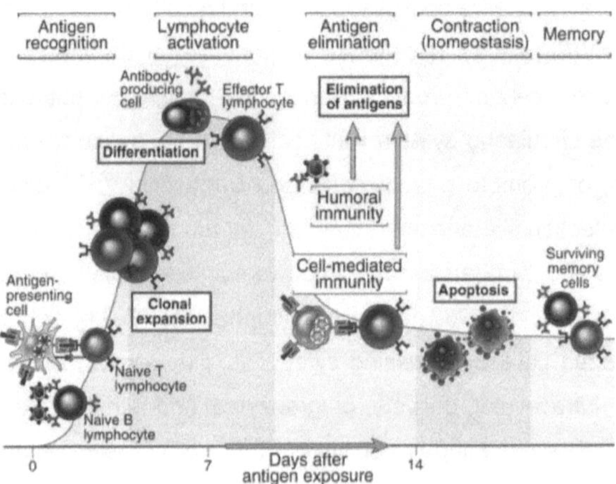

Figure 7. **Phases of an adaptive immune response.**
During an adaptive immune response, naïve lymphocytes recognize the antigen presented by antigen-presenting cells (APCs), and are activated, leading to their clonal expansion and their differentiation into specialized subsets. The effector phase declines, as activated lymphocytes die by apoptosis, and immune homeostasis is restored. Some lymphocytes acquire a long-lasting phenotype and are responsible for immune memory to the encountered pathogen. The y-axis represents an arbitrary measure of the magnitude of the response. These principles apply to humoral immunity (mediated by B lymphocytes) and cell-mediated immunity (mediated by T lymphocytes).
From: Abbas AK, Lichtman AH: Basic immunology: functions and disorders of the immune system, updated edition, ed 3, Philadelphia, 2011, Saunders.

protects the body from re-infection. Secondary immune responses are therefore faster and more efficient at eliminating pathogens. Vaccination draws on the establishment of immune memory upon encountering a pathogen, while providing a nearly symptom-free first encounter[4,56,57].

Certain immune cells lie at the crossroads between innate and adaptive immunity (**Figure 3**). It is the case of innate lymphoid cells (ILC), natural killer T (NKT) cells, and γδ T cells. These unconventional T cells show rapid effector responses and limited antigen specificities[58]. These cells reside in tissues, most prominently at barrier sites, like mucosal surfaces and skin, and play a key role in tissular immunity.

1.3. T cell development

To detect, eliminate, and remember a large number of pathogens, the adaptive immune system must be able to distinguish between an infinite number of, sometimes very closely related, antigens. T and B cell receptors must therefore be produced in a huge variety of configurations, one for each antigen that might be encountered. Repertoire diversity is generated by a random somatic recombination of gene fragments[36,37]. The random nature of this process is concomitant with the rise of autoreactive receptors. In order to prevent autoimmunity and maintain central tolerance, a selection process aimed at eliminating autoreactive receptors occurs during cell differentiation. This selection occurs in primary lymphoid organs, where lymphocytes are generated. B cell selection occurs in the bone marrow. T cell precursors migrate from the bone marrow to the thymus, where T cell selection occurs. This next part will focus on T cell selection. As will be explained below, the thymus selects the useful, neglects the useless, and destroys the harmful.

Upon entering the thymus, early thymocyte progenitors, which eventually differentiate into T cells, initiate TCR rearrangement[36,37] (**Figure 8**). The TCR is a heterodimeric glycoprotein, homologous to immunoglobulins, and made of an α chain and a β chain, covalently linked by disulfide bonds (**Figure 6**). TCR chains contain a variable and a constant region. The variable region is assembled at the genetic level by the somatic recombination of V and J gene fragments in the α chain gene locus, and V, D, and J fragments, in the β chain gene locus by Rag endonucleases. The TCR is associated to

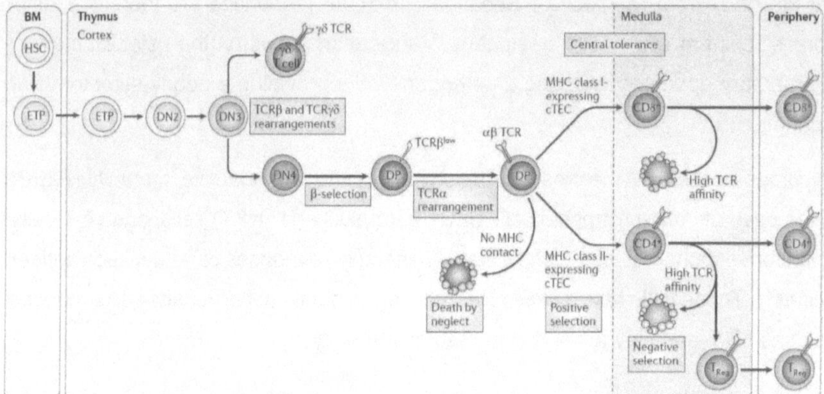

Figure 8. **T cell development and thymic selection.**
Hematopoietic stem cells (HSCs) from the bone marrow give rise to early thymic progenitors (ETPs). ETPs subsequently become double negative (DN) cells: they express neither the TCR, nor co-receptors CD4 and CD8. DN cells begin rearranging the TCR genes, acquire CD4 and CD8 expression, and become double positive cells (DP) upon rearrangement of the TCRβ chain. DP cells then rearrange of the TCR α-chain locus and express the αβ TCR. Cells with functional TCRs are positively selected, while cells with dysfunctional TCRs die by neglect. During positive selection, cells with TCRs that bind to MHC class I molecules retain expression of CD8, while those that bind to MHC class II molecules retain CD4. During negative selection, cells presenting a high affinity TCR for self-antigens are deleted or acquire a regulatory phenotype. cTEC, cortical thymic epithelial cell.
From: Miller, Jacques F. A. P. "The Golden Anniversary of the Thymus." *Nature Reviews Immunology* 11, no. 7 (July 2011): 489–95. https://doi.org/10.1038/nri2993.

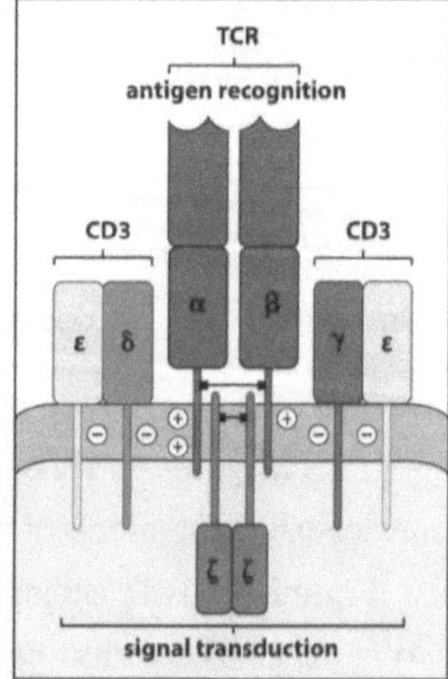

Figure 9. **TCR complex.**
The TCR is a heterodimeric glycoprotein, homologous to immunoglobulins, and made of an α chain and a β chain, covalently linked by disulfide bonds. It is associated to an invariant protein complex, the CD3 complex, which allows TCR signal transduction.
From: Parham, Peter. *The Immune System, 3rd Edition.* 3rd edition. London ; New York: Garland Science, 2009.

an invariant protein complex, the CD3 complex, which allows TCR signal transduction (**Figure 9**).

The different stages of T cell development are distinguished based on the expression of coreceptors CD4 and CD8, and of the TCR complex[59]. During the double negative (DN) stage, thymocytes are negative for both CD4 and CD8 expression and rearrange their TCRβ chain to form pre-TCR complexes (**Figure 8**). Signaling through pre-TCR complexes induces the expression of both CD4 and CD8. Thymocytes thus transition to the double positive (DP) stage and start rearranging their TCRα chain. Henceforth, CD4[+] CD8[+] DP thymocytes express αβ TCR complexes and can recognize self-peptides in the thymus[60,61].

During the positive selection stage, thymocytes with functional αβ TCR recognize self-peptide-MHC complexes presented by cortical thymic epithelial cells and are positively selected to continue the maturation process[60,61] (**Figure 8**). Other thymocytes do not receive survival signals and die by neglect. There are two classes of MHC molecules, which present antigen of distinct origins[41,42] (**Figure 10**). MHC class I molecules are expressed by all nucleated cells and present peptides of intracellular origin, while MHC class II are expressed exclusively by antigen presenting cells and present antigens of extracellular origin, internalized though endocytosis or phagocytosis. The coreceptors CD8 and CD4 bind to the constant regions of MHC class I and class II, respectively, thus facilitating TCR recognition of peptides presented by MHC molecules. After positive selection, thymocytes commit to CD4 or CD8, depending on the class of the MHC molecule that was successfully recognized by the TCR, and mature into single positive (SP) cells[60,61] (**Figure 8**). This MHC restriction accounts for the functional differences between CD4 and CD8 T cell subsets, CD4 T cells being tasked with the recognition of extracellular pathogens, while CD8 T cells recognize intracellular pathogens.

By this stage, positive thymic selection has "selected the useful" and "neglected the useless". However, cells with autoreactive TCR remain and must be eliminated to prevent autoimmunity. The expression of *Aire* by medullary thymic epithelial cells, or APCs, allows them to present ubiquitous and tissue-specific antigens, driving the negative selection of autoreactive T cells[62,63]. During the negative selection, cells with a TCR that recognizes

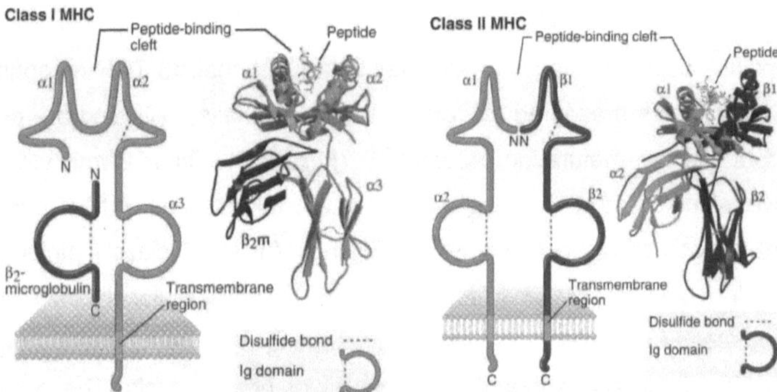

Figure 10. **Structure of the MHC molecules.**
MHC molecules are transmembrane presenting molecules made up of two chains that fold together and form a long cleft in which the peptide nests. MHC class I molecules are expressed by all nucleated cells and present peptides of intracellular origin, while MHC class II are expressed exclusively by antigen presenting cells and present antigens of extracellular origin, internalized though endocytosis or phagocytosis.
From: Abbas, Abul K, Andrew H Lichtman, and Shiv Pillai. Cellular and Molecular Immunology. Philadelphia: Elsevier/Saunders, 2012.

self-antigens with high or intermediate affinity are either clonally deleted, or committed to the regulatory T cell lineage, thus ensuring self-tolerance[64] (**Figure 8**). Mature naïve T cells then exit the thymus and reach secondary lymphoid organs, via the blood vessels, where immune responses are initiated.

1.4. CD4 T cell specialization

Helper T cells (Th cells) are tasked with guiding effector immune cells and orchestrating the immune response to efficiently eliminate different types of threats. To do so, Th cells need to integrate distinct environmental signals and adapt their phenotype accordingly.

1.4.1. General concepts

Th cell activation requires a combination of three distinct signals (**Figure 11**). The first signal depends on TCR recognition of a peptide-MHC class II complex, presented by an APC, and ensures response specificity. Antigen recognition induces a cascade of tyrosine phosphorylations, activating downstream pathways which facilitate differentiation, proliferation, and cell migration[65]. TCR activation alone leads to cell inactivation, unless a second signal is also provided by APCs[66,67]. PAMP recognition leads to APC maturation and expression of positive co-stimulation receptors, like CD80/86, which activate CD28 in T cells, and potentiate T cell signaling pathways, protecting cells from anergy or early apoptosis[68,69]. Cell co-stimulation acts as a confirmation signal, linking the development of an adaptive immune response to the previous recognition of an infectious event. Lastly, cytokines expressed by immune or non-immune cells allow helper T cells to integrate environmental signals and influence their differentiation fate toward a determined specialized subset[54].

1.4.2. Functional specialization

In order to optimize their performance, CD4 T cells differentiate into specialized subsets in response to environmental signals (**Figure 12**). This specialization allows helper T cells to activate immune cells which are able to respond to a specific pathogen. This is crucial as an inappropriate CD4 T cell response can lead to a lapse in pathogen control or in pathology development[70,71]. Each Th cell subset is induced by specific polarizing cytokines, often produced by DCs[54]. Cytokine signaling induces the expression of lineage-

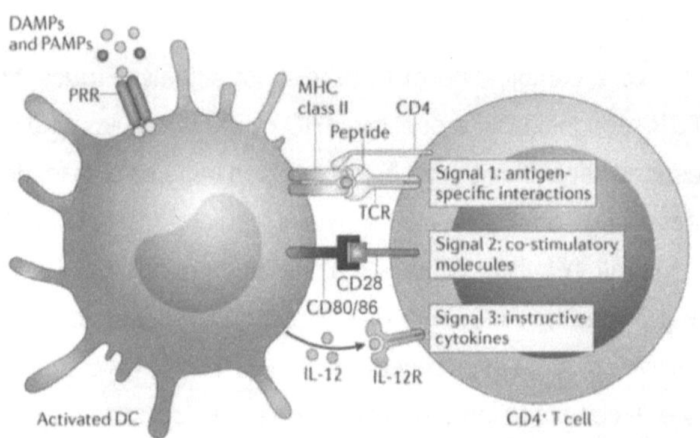

Figure 11. **T cell activation requires three signals.**
In order for a T cell to be activated, three signals have to be provided. Signal 1 depends on TCR recognition of a peptide-MHC class II complex, presented by an APC, and ensures response specificity. Signal 2 depends on co-stimulatory signals. It is provided by mature APCs and acts as a confirmation signal, linking the development of an adaptive immune response to the previous recognition of an infectious event. Signal 3 depends on cytokine signaling, allowing the integration of environmental signals, and leading to the specialization of T cells.
Adapted from: Kambayashi, Taku, and Terri M. Laufer. "Atypical MHC Class II-Expressing Antigen-Presenting Cells: Can Anything Replace a Dendritic Cell?" *Nature Reviews Immunology* 14, no. 11 (November 2014): 719–30.

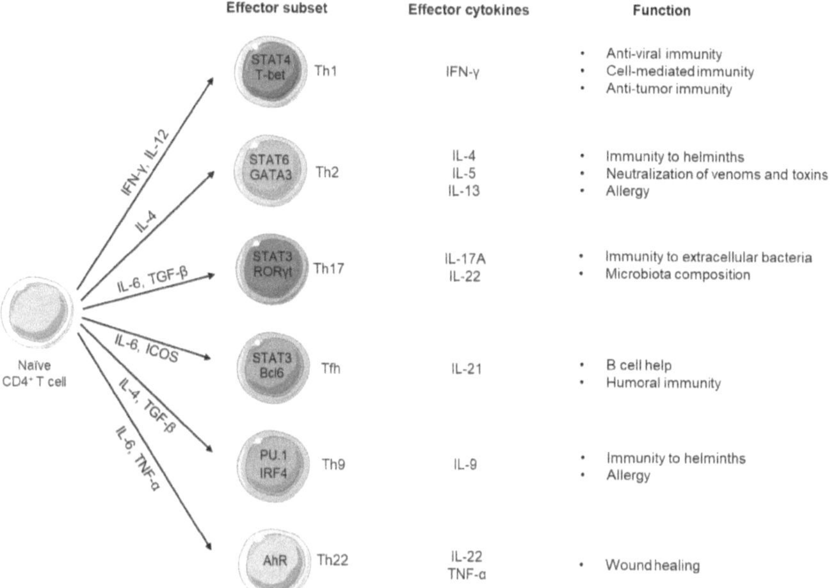

Figure 12. **Specialized helper T cell subsets.**
In order to optimize their performance, CD4 T cells differentiate into specialized subsets in response to environmental signals.

defining transcription factors, called 'master' transcription factors. These master transcription factors consequently initiate transcriptional programs which lead to the expression of specific chemokine receptors and effector cytokines, and allow the cells to tailor their effector functions to the environment. Master transcription factors often antagonize each other's expression, thus reinforcing differentiation into a given subset[54]. However, certain subsets can show plasticity[72–74], which is highly dependent on environmental conditions.

1.4.2.1. Th1 / Th2 subsets

Initial work into CD4 T cell function established a dichotomy between two functional subsets characterized by different cytokine expression profiles. Th1 cells were identified by their production of IFN-γ, whereas Th2 cells were characterized by the production of IL-4, IL-5, and IL-13, thus establishing the Th1/Th2 paradigm[75]. This is accentuated by the fact that Th1 cells can inhibit Th2 cytokine production and differentiation by producing IFN-γ[76–78], and that subset-specific loci were shown to carry repressive H3K27me3 histone modifications in opposing cell subsets (*Tbx21* and *Ifng* in Th2 cells and *Gata3* and *Il4* in Th1 cells)[79].

a. Th1

Several cytokines are involved in the differentiation of Th1 cells. IFN-γ produced by innate immune cells activates STAT1 signaling, which in turn promotes the expression of T-bet[80], the master transcription factor of Th1 cells. Th1 differentiation is amplified by IL-12[81], which acts through STAT4, and promotes IFN-γ production, thus creating a feed forward cycle to amplify the Th1 response. In addition to T-bet, transcription factors like Runx3 and Hlx are important for optimal Th1 function[82,83].

Th1 cells are the principal regulators of type 1 immunity against intracellular pathogens, such as viral infections and intracellular bacteria, as well as tumors[84]. IFN-γ is a pro-inflammatory cytokine which promotes pathogen clearance, by potentiating the phagocytic abilities of APCs as well as their ability to activate CD8 T cells, and by directly stimulating NK cells and CD8 T cells. In addition to IFN-γ, Th1 cells produce IL-2, as well as lymphotoxin, and TNF-α, which contribute to microbial defense. Th1 cells express chemokine receptor CXCR3, which allows them to respond to chemokines CXCL9 and

CXCL10 and traffic into inflamed tissues[85–87]. Th1 cells also stimulate antibody class switching to IgG2a antibodies, which optimize clearance of viruses and extracellular bacteria[88,89].

Increased susceptibility to infections in humans deficient for IFN-γ, IL-12, or STAT1 supports the essential role of Th1 cells in the defense against intracellular infections[90–92]. However, exacerbated Th1 responses have deleterious consequences and have been involved in the pathogenesis of multiple sclerosis and diabetes.

b. Th2

Th2 cells are involved in type 2 immune responses against extracellular parasites, particularly parasitic helminths[93,94], and with the neutralization of venoms and toxins[95].

IL-4 produced by mast cells and basophils interacts with its receptor expressed at the surface of naïve CD4 T cells, signals via STAT6, and induces the expression of GATA3, the master regulator of Th2 differentiation. Diminished TCR stimulation as well as CD28 co-stimulation are also important to promote Th2 differentiation[96]. IL-4 production by mature Th2 cells promotes Th2 differentiation in naïve T cells, thus inducing a positive feedback loop. Thymic stromal lymphopoietin (TSLP) is a cytokine expressed mainly by epithelial cells and epidermal keratinocytes, and is involved in the differentiation of Th2 cells. In presence of TCR stimulation, TSLP promotes Th2 differentiation by upregulating IL-4 expression in T cells, which will in turn lead to the upregulation of TSLP expression, resulting in a positive feedback loop[97,98].

The signature cytokines associated with a Th2 response are IL-4, IL-5, and IL-13. IL-4 stimulates B cell class-switching to IgE. IgE crosslinks FcεRI on basophil and mast cells leading to their degranulation, production of histamine and serotonin as well as IL-4, IL-13, and TNF-α[99]. IL-5 stimulates eosinophil production as well as activation and chemotaxis of eosinophils and basophils to affected tissue. IL-13 induces mucus production in epithelial cells, increases smooth muscle contractibility in the gut and lungs, and thus helps with parasite expulsion. Th2 cells can also produce IL-10, IL-25, and amphiregulin and express chemokine receptor CCR4 with allow them to migrate to the lung and airways[100].

Th2 cells are also involved in the development of allergic diseases such as atopic dermatitis, allergic rhinitis, and asthma[94]. In addition, aberrant Th2 responses to infection can lead to severe disease following the infection with pathogens that require Th1-mediated responses for their elimination. This is the case of *Leishmania major*, which is cleared by Th1 responses, but which leads to severe cutaneous and systemic disease in the context of Th2 responses[70]. Similarly, in lepromatous leprosy, Th2 responses are ineffective at resolving *Mycobacterium leprae* infection and cause severe disease, while tuberculoid leprosy is associated with strong Th1 responses and leads to a less destructive disease[71].

1.4.2.2.   Th17

The discovery of a third subset of CD4 T cell challenged the Th1/Th2 paradigm. Th17 cells were identified in 2005[101]. They are characterized by the expression of IL-17A and IL-17F, but also express IL-22 and TNF-α[102]. Th17 cells are crucial for the protection against extracellular bacteria and fungi, positioning them as important actors in establishing intestinal homeostasis and controlling microbiota composition. Th17 cells express chemokine receptor CCR6, which allows them to migrate to mucosal environments, and are involved in the activation and migration of neutrophils, promotion of anti-microbial peptide (AMP) production by epithelial cells, as well as the maintenance and repair of the intestinal epithelium[102]. IL-17A and IL-17F induce IL-6, GM-CSF, and G-CSF expression, which promote neutrophil-mediated inflammation[102]. IL-22 induces AMP production and mucus secretion[103,104].

Naïve T cells differentiate into Th17 cells when stimulated with TGF-β and IL-6 and start expressing RORγt, the master regulator of Th17 differentiation[105]. While IL-6 directly upregulates RORγt expression through STAT3 signaling, TGF-β has an indirect effect, and inhibits the expression of STAT4 and GATA3, necessary for Th1 and Th2 differentiation, respectively[102,106]. This is important as Th1- and Th2-inducing factors suppress Th17 differentiation. Indeed, IFN-γ represses Th17 cell differentiation in a T-bet and STAT1-dependent manner[107–109]. IL-23 is not required for the differentiation of Th17 cells but is necessary for their maintenance[110]. IL-1β also promotes Th17 differentiation and increases responsiveness to IL-23[111,112].

Th17 cells are also known to play a role in the pathogenesis of autoimmune disorders, like multiple sclerosis, rheumatoid arthritis, and psoriasis[106]. Th17 cells can be grouped into two types. While non-pathogenic Th17 cells are induced by TGF-β signalling, the induction of pathogenic Th17 cells is linked to the secretion of IL-6, IL-1β, and IL-23 by endothelial and epithelial cells. Pathogenic Th17 cells have a Th1/Th17 phenotype, and start expressing T-bet and IFN-γ[113], as well as TGF-β3[114], in addition to the expression of RORγt and IL-17. Pathogenic Th17 cells, but not TGF-β-induced Th17 cells are involved in the development of experimental autoimmune encephalomyelitis (EAE), a murine model of multiple sclerosis[112].

TGF-β is important for the differentiation of both Th17 cells and peripherally induced regulatory T cells (pTreg)[115,116]. Regulatory T cells (Treg) are suppressive CD4 T cells, that originate in the thymus or in the periphery. The origin, function, and differentiation mechanisms of regulatory T cells will be detailed at length in the following sections. The balance between Th17 cells and pTreg cells is therefore influenced by environmental cues, like IL-6 availability[117]. This developmental co-dependence is accentuated by the plasticity between Th17 cells and Treg cells. Indeed, Th17 cells have been shown to transdifferentiate into regulatory T cells and contribute to the resolution of inflammation[118].

1.4.2.3. Tfh

Follicular helper T cells, or Tfh cells, were initially identified in human tonsils and are located in the follicular region of secondary lymphoid organs[119–121]. Tfh cells are specialized in providing B cell help, promoting antibody production, somatic hypermutation, and class switching, and support the generation of long-lasting humoral immunity[122,123]. Tfh cells are characterized by the expression of surface markers ICOS and PD-1, and chemokine receptor CXCR5, which allows them to migrate to B cell zones[119,120], and the production of IL-21.

The cytokines IL-6 and IL-21 promote Tfh cell differentiation[124–126]. Both IL-6 and IL-21 signal through STAT3 and STAT3-deficient CD4 T cells demonstrate impaired differentiation of Tfh cells[125]. The commitment to the Tfh phenotype starts in the T cell zone of secondary lymphoid organs, where DCs provide ICOS co-stimulation to naïve T cells, thus inducing the expression of transcriptional repressor Bcl6[127], the master regulator of Tfh differentiation[128,129]. Bcl6 upregulates CXCR5 expression on the surface of pre-Tfh cells, allowing them to migrate to the border between the B and T cell area[124]. Maintenance of the Tfh phenotype is dependent on the interaction with B cells. Germinal center Tfh cells are fully polarized Tfh cells which express the highest levels of Bcl6 and CXCR5. Tfh cells provide B cell help through the expression of co-stimulation molecules, like CD40L and ICOS, but also the secretion of IL-21, essential for B cell survival[122,123]. Tfh cells induce the development of germinal centers, where B cells undergo clonal proliferation as well as subsequent affinity maturation and antibody class-switching. This process results in the increased production of antibodies with high antigen affinity. Tfh cells also provide the signals required for memory B cell differentiation, positioning them as major contributors to the development of efficient vaccinal responses.

Recent literature argues that Tfh cells are not a distinct T cell subset, but rather represent a distinct activation state within any given effector subset[55]. Indeed, although most Tfh cells produce IL-21, some of them rely on the expression of transcription factors T-bet and GATA3 to induce IFN-γ and IL-4 expression, which favor antibody class switching to IgG2a or IgE, respectively[130].

1.4.2.4.   Others

Recently, additional CD4 T cell subsets have been identified, but their respective master regulators have not yet been identified. For example, Th9 cells have been implicated in the protection against helminths, but also play a role in allergic airway inflammation and the pathogenesis of autoimmune diseases, like EAE[131]. Th9 cells are characterized by their IL-9, IL-10, and IL-21 production. IL-9 contributes to mucus expression and mast cell development[132,133]. Naïve T cells differentiate into Th9 cells after stimulation with IL-4 and TGF-β, and rely on the expression of transcription factors PU.1, IRF4, as well as GATA3[131]. It is unclear if Th9 cells are a distinct lineage or a specialized subset of Th2 cells. Indeed, Th9 cells and Th2 cells rely on similar factors for their differentiation and Th2 cells have been shown to be reprogrammed into IL-9-producing Th9 cells after stimulation with TGF-β[134]. Recently, Th9 cells have been shown to promote protective responses against solid tumor cancers[135].

Th22 cells differentiate in the presence of IL-6 and TNF-α and are characterized by their expression of IL-22[136–139]. IL-22 production is linked to the expression of aryl hydrocarbon receptor (AhR) and promotes antimicrobial responses, inhibits epithelial cell damage in the intestine and the lungs, and stimulates epithelial repair[140]. IL-22 is also expressed by Th17 cells, but Th22 cells are unique in that they express IL-22 in the absence of IFN-γ, IL-4, or IL-17, suggesting that they form a separate lineage[136–139]. Th22 cells are also characterized by the expression of chemokine receptors CCR4, CCR6, and CCR10, allowing them to home to the skin[141]. Th22 cells have also been shown to promote gastrointestinal tumors and play a role in the pathogenesis of psoriasis[142,143].

1.4.3. Regulatory T cells

CD4 T cells play a dual role in immunity. Indeed, while conventional CD4 T cells promote inflammatory immune responses, regulatory CD4 T cells are tasked with controlling aberrant or excessive immune responses, in order to maintain immune homeostasis, thus ensuring optimal pathogen protection as well as organ function.

1.4.3.1. General concepts

Regulatory T cells (Treg) are a subset of suppressive CD4 T cells. They regulate excessive immune responses using a variety of suppressive mechanisms and are instrumental in preventing inflammation and autoimmunity[144].

A population of 'suppressor T cells' was initially discovered in 1970[145]. Because of the lack of reliable markers to differentiate between suppressor and other T cells, the study of suppressor T cells was quickly dropped[146]. In 1995, a subset of suppressive T cells was shown to express high levels of CD25, the α chain of the IL-2 receptor, which helped identify and further study this cell population[147,148]. In 2003, transcription factor Forkhead box P3 (Foxp3) was identified as the master transcription factor of regulatory T cells[149–151]. Ectopic expression of Foxp3 by retroviral transduction is sufficient to induce regulatory activity in T cells[150,151]. Foxp3 is essential for Treg differentiation, maintenance, and function[152]. Indeed, mutations in Foxp3 lead to the development of autoimmunity, identified as the scurfy phenotype in mice[153] and the immunodysregulation, polyendrocrinopathy, enteropathy, X-linked (IPEX) syndrome in humans[154,155]. While Foxp3 is a reliable marker for detecting functional regulatory T cells in mice, this is not the case in humans where Foxp3 expression does not correlate with Treg suppressive function, and additional markers, like CD25 and CD127 have to be used[156].

1.4.3.2. Structure of the Foxp3 protein

Transcription factor Foxp3 belongs to the forkhead–winged-helix family. Foxp3 is a 47-kDa protein composed of 429 amino acids in mice and 431 amino acids in humans[157,158] (**Figure 13**). It comprises three functionally important domains: an N-terminal domain, a zinc finger and leucine zipper-containing central domain, and a C-terminal forkhead domain[144,152]. The forkhead domain is necessary for DNA binding[159], even though its deletion does not lead to the complete loss of Foxp3 transcriptional activity[160]. The central domain zinger finger and leucine zipper domain are required for homodimerization of Foxp3 and protein complex formation[161–164]. The ability of Foxp3 to bind to promoter sequences relies on its oligomerization status, which also relies on the forkhead domain[163]. The N-terminal domain is a proline-rich repressor domain, which helps protein interaction, but is also essential to the repression of target genes[164].

1.4.3.3. Gene regulation by Foxp3

Foxp3 has been shown to form a 400-800kDa transcriptional complex made of over 300 protein co-factors[165]. This complex also allows chromatin remodeling by interacting with histone remodelling proteins, like HDAC7 or HDAC9 and KAT5[165,166]. Foxp3 controls the transcription of over 2,800 genes, including 700-1400 Treg signature genes[167,168]. Foxp3 binds directly to only 6-10% of the genes it has transcriptional control over and most of the binding occurs indirectly via Foxp3 co-factors[167]. Foxp3 has been shown to act as both a transcriptional activator and a transcriptional repressor[167,168]. Foxp3 relies heavily on other transcription factors, such as Eos, c-Rel, GATA3, RORγt, and HIF-1α for its binding and transcriptional control over the Treg transcriptional signature[165,169,170]. Foxp3 and its partners form a regulatory network with positive and negative feedback loops, which ultimately lead to the acquisition of the full Treg transcription signature, particularly the members of the 'quintet', Eos, IRF4, GATA-1, Lef1 and Satb1[165,170]. Foxp3 also interacts with transcription factors nuclear factor of activated T cells (NFAT) and Runt-related transcription factor 1 (Runx1), both key regulators of T cell activation and anergy, which bind to the promoter regions of Foxp3-regulated genes[171,172]. In conventional T cells,

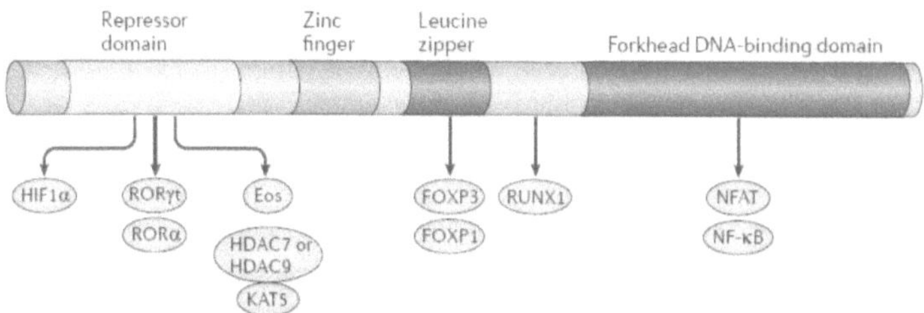

Figure 13. **Protein structure of Forkhead box protein 3 (Foxp3)**.
Forkhead box P3 (Foxp3) has four functional domains: the repressor domain, in the N-terminal portion; the zinc finger (ZnF) and leucine zipper (Zip) domains, in the central portion; and the forkhead domain, in the C terminal portion.
Foxp3 interacts with different binding partners. HDAC, histone deacetylase; HIF1α, hypoxia-inducible factor 1α; KAT5, histone acetyltransferase KAT5; NFAT, nuclear factor of activated T cells; NF-κB, nuclear factor-κB; RORα, retinoic acid receptor-related orphan receptor-α; RORγt, retinoic acid receptor-related orphan receptor-γt; RUNX1, runt-related transcription factor 1.
From: Ramsdell, Fred, and Steven F. Ziegler. "FOXP3 and Scurfy: How It All Began." *Nature Reviews Immunology* 14, no. 5 (May 2014): 343–49.

NFAT forms complexes with activator protein-1 (AP-1) and activates the transcription of activation-associated genes. Foxp3 competes with AP-1 for NFAT binding and thus interferes with T cell activation[171]. By interacting with NFAT and nuclear factor kappa B (NF-κB), Foxp3 also blocks the expression of IL-2, IL-4, and IFN-γ[173]. Foxp3 also upregulates typical Treg surface markers, such as CD25, CTLA-4, GITR and CD103 expression[150], which illustrates its role as a transcriptional activator.

1.4.3.4.   Structure of the *Foxp3* gene

The *Foxp3* gene is carried by the X chromosome, measures 21 kb and comprises 11 exons[144] (**Figure 14**). The gene sequence is highly conserved between human and mouse, especially at the exon-intron interfaces[153,155]. In response to TCR signalling and co-stimulation pathways, transcription factors such as NFAT, AP-1, FOXO1 and 3, or CREB-ATF1 complexes bind to the *Foxp3* promoter[174–176]. The *Foxp3* promoter shows a low transactivating potential and relies on conserved enhancer regions for transcription initiation and maintenance[174]. The *Foxp3* locus contains four conserved non-coding sequences (CNS), which serve as binding sites for several transcription factors[177] (**Figure 14**). These cis-regulatory elements intervene to varying degrees in the development of Treg cells, depending on their developmental origin. This will be explained further in the following section. CNS1 is required for the induction of Foxp3 expression in peripheral Treg cells, but dispensable for thymic Treg differentiation. Oppositely, CNS3 is required for the expression of Foxp3 in thymic Treg cells, but not peripheral Treg cells. CNS2, also known as Treg cell-specific demethylated region (TSDR), contains CpG islands that are highly demethylated in functional Treg cells. CNS2 does not affect Foxp3 induction but rather controls the maintenance of Foxp3 expression. Recently, a super-enhancer named CNS0 was discovered upstream of the *Foxp3* promoter. The binding of Satb1 at CNS0 initiates chromatin looping, bringing closer other regulatory regions[178].

1.4.3.5.   *Foxp3* gene transcription and Treg origin

Regulatory T cells can be distinguished based on their origin[†]. While thymic Treg cells (tTreg) differentiate in the thymus, peripheral Treg cells (pTreg) are induced in the

---

[†] Thymic Tregs (tTreg) and peripheral Tregs (pTreg) are part of the new nomenclature suggested by Abbas et al[179], and were referred to in previous publications as 'natural Tregs' (nTreg) and 'induced Tregs' (iTreg), respectively.

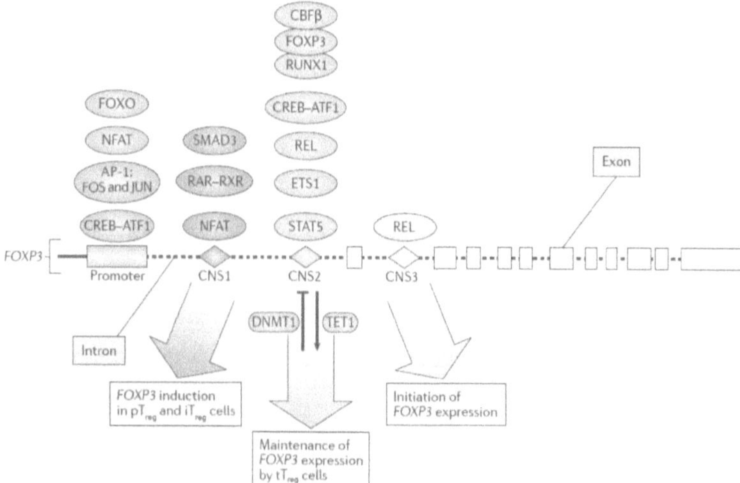

Figure 14. **Regulatory elements within the *Foxp3* gene locus.**
In addition to its promoter, the *Foxp3* locus comprises several regulatory elements. Conserved non-coding sequence 1 (CNS1) is involved in Foxp3 induction in pTreg and iTreg cells. CNS2 contains demethylated CpG residues and is involved in the stability of Foxp3 expression. The methylation status of the CNS2 is regulated by enzymes such as DNA methyltransferase 1 (DNMT1) and ten-eleven translocation 1 (TET1). CNS3 is involved in the initiation of *Foxp3* expression in tTreg cells. Several transcription factors bind to the *Foxp3* locus. Transcription factors that bind to the promoter, CNS1, CNS2 and CNS3 regions of *Foxp3* are shown in blue, red, green and yellow, respectively. ATF1, activating transcription factor 1; CBFβ, core-binding factor subunit-β; CREB, cAMP-responsive element-binding protein; FOXO, forkhead box protein O; iTreg, *in vitro*-induced regulatory T cell; NFAT, nuclear factor of activated T cells; pTreg, peripherally derived regulatory T cell; RAR, retinoic acid receptor; RUNX1, Runt-related transcription factor 1; RXR, retinoid X receptor; STAT5, signal transducer and activator of transcription 5; tTreg thymus-derived regulatory T cell.
From: Lu, Ling, Joseph Barbi, and Fan Pan. "The Regulation of Immune Tolerance by FOXP3." *Nature Reviews Immunology* 17, no. 11 (November 2017): 703–17.

periphery (**Figure 15**).

tTreg cells arise in the thymus during negative selection[64]. At this stage, T cells with self-reactive TCRs are selected based on their level of affinity to self-antigens. Nur77 is induced proportionally to the intensity of the TCR[180], binds to the proximal promoter of Foxp3 and is thus thought to translate TCR affinity into a T cell fate decision. T cells with TCRs of no or low affinity for self-antigens are negatively selected, while TCRs of intermediate affinity trigger Foxp3 induction and lead to the acquisition of a suppressive phenotype[180–182]. TCR stimulation triggers the activation of c-Rel and other NF-κB family members, which bind to the CNS3 to initiate Foxp3 transcription[177,183,184]. The stability of Treg cells is epigenetically regulated and relies on the demethylation of CNS2, which is initiated during early stages of thymic development[177]. The demethylation of CNS2 allows Foxp3 itself to bind to CNS2, with the help of Runx1, and results in stable Foxp3 expression[185]. IL-2 signalling is necessary for the maintenance of Foxp3 expression. It activates STAT5, which binds to CNS2 and contributes to its demethylation[186,187]. Interestingly, IL-6 activates STAT3, which has been shown to compete with STAT5 for the binding to STAT binding sites within the *Foxp3* gene, and therefore destabilize Foxp3 expression[188]. Unlike stable Treg cells, cells that transiently upregulate Foxp3 show incomplete demethylation of the CNS2 CpG residues.

On the other hand, pTreg cells arise upon induction of Foxp3 expression in naïve CD4 T cells in the periphery, after exposure to TGF-β, retinoic acid, or short chain fatty acids (SCFA)[189]. TGF-β signalling induces the binding of SMAD3 on CNS1, which induces Foxp3 expression in naïve CD4 T cells[190]. Retinoic acid, or all-*trans*-retinoic acid, is a metabolite of vitamin A, which binds to intracellular nuclear receptors called retinoic acid receptors (RAR). The heterodimers formed by RAR and retinoid X receptors (RXR) bind to the CNS1, thus contributing to the induction of Foxp3[191,192]. Retinoic acid also influences Foxp3 expression by boosting the TGF-β-induced phosphorylation of SMAD3[193]. Moreover, retinoic acid induces histone H4 acetylation at the *Foxp3* locus, which opens up the chromatin and favors *Foxp3* transcription[194]. Similarly, small chain fatty acids, such as butyrate and propionate, promote pTreg generation in a CNS1-dependent manner and by enhancing the histone acetylation of the *Foxp3* locus[195,196].

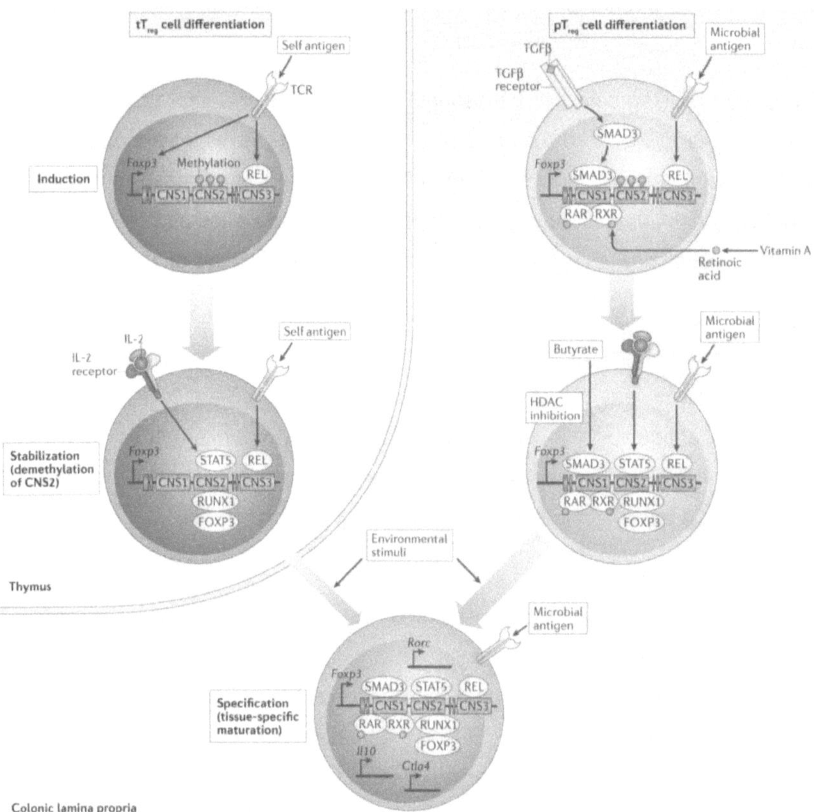

Figure 15. **Induction of tTreg cells and pTreg cells.**
tTreg cell differentiation is initiated by strong TCR signals induced by self antigens in the thymus. This induces nuclear entry of REL which binds to the *Foxp3* promoter and the CNS3, thus inducing *Foxp3* gene expression. pTreg cell differentiation occurs in the periphery, from naive CD4 T cells. pTreg differentiation is initiated by the recognition of

peripheral antigens in presence of TGF-β and retinoic acid, which induces the binding of REL to the CNS3, as well as the binding of SMAD3 and RAR–RXR complexes to the CNS1. Stable Treg lineages have a demethylated CNS2, where transcription factors STAT5, Runx1, and Foxp3 can bind, resulting in stable *Foxp3* expression. Butyrate inhibits histone deacetylases and contributes to the stabilization of the pTreg lineage. Treg cells subsequently migrate to the tissues and further undergo functional maturation in response to environmental stimuli, acquiring the expression of RORγt, IL-10 and CTLA-4 in the intestinal environment.

From: Tanoue, Takeshi, Koji Atarashi, and Kenya Honda. "Development and Maintenance of Intestinal Regulatory T Cells." *Nature Reviews Immunology* 16, no. 5 (May 2016): 295–309.

Treg cells can also be induced *in vitro* (iTreg) by activating naïve CD4 T cells in presence of TGF-β and IL-2[197]. While pTreg cells show a mostly demethylated CNS2[198], indicating a stable suppressive phenotype, iTreg cells lack the Treg cell–specific demethylation signature[199,200].

The origin of murine Treg cells can be distinguished based on the expression of the cell surface protein neuropilin 1 (Nrp1) and the nuclear protein Helios, which are constitutively expressed on tTreg cells but not on pTreg cells[201–203]. The use of these molecules as markers of Treg origin is not unambiguous. Indeed, Nrp1 and Helios can be upregulated in inflammatory conditions, thus hindering the accurate identification of tTreg and pTreg cells[204].

Treg cells of different origins have different TCR repertoires. Indeed, while the tTreg cell repertoire is biased toward the recognition of self-antigens, pTreg cells preferentially recognize foreign antigens[205–208]. Regardless of their repertoire differences, both tTreg and pTreg cells can be found in peripheral tissues and can undergo further functional and tissular specification, as will be explained in a following section.

1.4.3.6. Suppressive mechanisms

To control and resolve inflammatory responses, Treg cells are able to suppress cell proliferation, block effector function or induce apoptosis in effector cells. The suppressive capacity of Treg cells relies on a number of contact-dependent or -independent mechanisms (**Figure 16**). While some of these mechanisms are essential for the general suppressive function of Treg cells, other mechanisms are organ-specific and serve to control local immunity. In order to become activated and exert their suppressive activity, Treg cells require antigen-specific TCR engagement. However, once activated, Treg cells can exert bystander suppression in a non-antigen specific manner[209,210]. It is likely that while some suppression mechanisms used by Treg cells are specific to certain types of immune responses, their function can show certain levels of redundancy.

1.4.3.6.1. Suppressive cytokines

The major non-specific suppression mechanism used by Treg cells is the production of the anti-inflammatory cytokines TGF-β, IL-10, and IL-35. These cytokines exhibit a broad range of suppressive functions[211]. TGF-β is critical for the development of Treg cells but

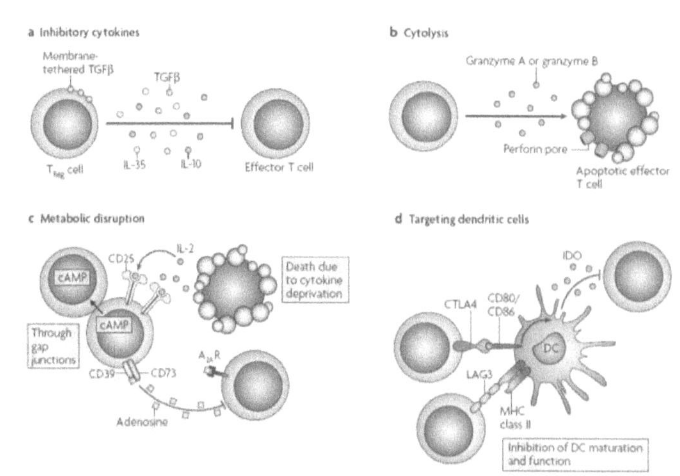

Figure 16. **Treg suppression mechanisms.**
Treg cells use different suppression mechanisms to control effector immune cells: a. Inhibitory cytokines, like IL-10, TGF-β, and IL-35; b. Cytolysis, in a granzyme-A/B-dependent and perforin-dependent mechanism; c. Metabolic disruption, by CD25-mediated IL-2 deprivation, cAMP-mediated inhibition, and CD39/CD73-generated adenosine generation; d. Inhibiting DC maturation and function by the expression of co-inhibitory molecules like LAG3 and CTLA-4, as well as the induction of IDO.
From: Vignali, Dario A. A., Lauren W. Collison, and Creg J. Workman. "How Regulatory T Cells Work." *Nature Reviews Immunology* 8, no. 7 (July 2008): 523–32.

is also used to mediate suppression[212]. TGF-β impedes T cell activation by inhibiting TCR signaling, but also opposes T effector cell differentiation, by inhibiting T-bet and GATA-3 expression[212]. TGF-β is produced in a latent form and has to be activated extracellularly to exert its functions[212]. Mature TGF-β is non-covalently associated to the latency associated peptide (LAP), forming the small latent complex. The LAP region folds around the mature cytokine, blocking access of TGF-β to its receptor[213]. In some cells, the small latent complex can associate with latent TGF-β binding protein (LTBP) and form the large latent complex[212]. Latent TGF-β can be enzymatically cleaved by serine proteases and metalloproteases, but can also be activated by integrins[212]. By binding to LAP, integrins αvβ6 and αvβ8 exert physical strain over the tethered latent complex and activate TGF-β. LTBP supports integrin-mediated TGF-β activation entirely through mechanically tethering the latent complex[214]. Activation of latent TGF-β by Treg cells is mainly mediated by glycoprotein A repetitions predominant (GARP). GARP is a transmembrane protein, covalently linked to LAP, like LTBP, that chaperones and orients TGF-β for binding and activation by integrin αvβ8[215]. The activation of latent TGF-β has also been shown to be mediated by Nrp1 in Treg cells[216].

While most immune cells are able to produce IL-10, Treg cells are particularly high producers of this anti-inflammatory cytokine. The production of IL-10 by Treg cells is especially important to regulate mucosal inflammatory responses. Indeed, Treg-specific IL-10 deficiency does not lead to the development of systemic autoimmunity in mice but does however cause spontaneous colitis and heightened skin inflammation[217], similarly to *Il10*$^{-/-}$ animals. IL-10 signaling is required for the control of Th17 cells at mucosal surfaces[218,219].

IL-35 is a member of the IL-12 family and is a heterodimer of IL-12p35 and Epstein-Barr virus-induced gene 3 (Ebi3). IL-35 is thought to be mainly secreted by Treg cells and to be essential for their optimal suppressive capacity[220]. Interestingly, IL-10 and IL-35 expression are segregated to different Treg subsets, but IL-10$^+$ IL-35$^+$ Treg cells can be found in the skin and intestine at steady state. In the context of cancer, IL-10 and IL-35 both drive inhibitory receptor expression on tumor-infiltrating lymphocytes, but IL-35 has a more pronounced effect on CD8 T cells[221].

1.4.3.6.2.   Metabolic disruption

IL-2 is a crucial factor in the proliferation of effector T cells. Treg cells express high levels of CD25, the high affinity receptor for IL-2. This high CD25 expression enables Treg cells to consume local IL-2 and thus starve actively proliferating effector T cells[222–225]. This is particularly the case with CD8 T cells[226], which show a higher dependence on IL-2 than CD4 T cells for their proliferation[227,228].

Another antigen non-specific mechanism involves the expression of ectoenzymes CD39/CD73 on the Treg surface. Treg cells express high levels of CD39 and CD73, which convert ATP/ADP to adenosine[229,230]. While ATP induces a pro-inflammatory milieu, adenosine activates A2a receptors on conventional T cells, increasing intracellular levels of cyclical AMP (cAMP), creating an anti-inflammatory milieu. cAMP inhibits TCR-mediated signaling by inhibiting Zap70 and AP-1 phosphorylation[231], and thus suppresses the proliferation of effector T cells. cAMP also inhibits antigen presentation by DCs. Treg cells contain high levels of cAMP[232] and are also able to deliver cAMP directly to effector cells through gap junctions[233].

1.4.3.6.3.   Co-inhibitory molecules

Antigen-specific suppression is mainly caused by the interaction of Treg cells with DCs via the expression of inhibitory co-stimulation molecules. This interaction leads to the acquisition of a tolerogenic phenotype by the DC or the inhibition of their antigen presenting capacity.

T cell activation depends on the provision of appropriate co-stimulatory signals, mainly mediated by the interaction of CD80/86, on the surface of DCs, with CD28, on T cells. Upon activation, Treg cells start expressing high levels of the co-inhibitory receptor cytotoxic T-lymphocyte-associated protein 4 (CTLA-4, CD152), which exhibits a higher affinity for CD80/86 than CD28[234], thus preventing potent T cell activation by limiting DC co-stimulation[235]. Treg cells are also able to remove antigen-MHC II complexes and CD80/86 from the DC surface by trans-endocytosis[236–238]. CTLA-4 stimulation of DCs also increases their expression of IDO, an enzyme involved in tryptophan catabolism, which leads to the production of pro-apoptotic metabolites and the suppression of effector T cells[239,240]. While the deletion of IL-10 expression in Treg cells only has local effects, deletion of CTLA-4 leads to the development of systemic immunity[217,241].

Other co-inhibitory molecules are expressed by Treg cells. Lymphocyte activation gene-3 (LAG-3, CD223) is structurally similar to CD4 and binds to MHC II with higher affinity than CD4, leading to the inhibition of TCR signaling and a reduced T cell proliferation and cytokine production[242]. Programmed cell death protein 1 (PD1, CD279) expression on Treg cells generates tolerogenic DCs but is also required for the maintenance of the Treg phenotype[243]. Tim-3 and TIGIT are also upregulated by Treg cells and have been involved in their suppressive function[242].

1.4.3.6.4.   Cytolysis

Similarly to CD8 T cells and NK cells, which are often thought of as the prototypical cytotoxic immune cell subsets[244], Treg cells exhibit cytotoxic properties and are able to induce apoptosis in target cells using several molecular mechanisms. Like CD8 T cells and NK cells, Treg cells show cytotoxicity via the secretion of granzyme and perforin[245,246]. Perforin is a pore forming protein that allows the delivery of granzyme A and B into the target cell. Granzymes are serine proteases that induce programmed cell death by activating caspases, compromising mitochondrial integrity, and favoring the release of pro-apoptotic mediators[244]. Treg cells have also been shown to induce cell apoptosis in a TRAIL-dependent manner. Activated Treg cells acquire the expression of TRAIL, which can bind to its ligand, DR5, on the surface of target cells, and induce apoptosis by activating caspase 8[247,248].

1.4.3.7.   Foxp3-negative suppressive T cell subsets

While Foxp3-expressing T cells constitute the main suppressive T cell subset, other suppressive T cell subsets have been discovered. These suppressive subsets do not express Foxp3, but nevertheless share suppression mechanisms with Treg cells.

Type 1 regulatory T cells (Tr1) can be induced both *in vivo* and *in vitro* upon chronic activation of CD4 T cells in presence of IL-10[249]. Tr1 cells have been identified as CD4+ CD49b+ LAG-3+ CD226+[250] and exert their immunosuppressive phenotype by producing high levels of IL-10[249,251]. The differentiation of Tr1 cells has also been shown to occur in presence of IL-27[252–254]. IL-27 is a heterodimeric cytokine produced by a variety of myeloid cells in response to TLR stimulation[255]. IL-27 signalling activates STAT3, which drives Blimp-1 and subsequent IL-10 expression, in an Egr2-dependent manner[256]. IL-27 also induces the expression of AhR and c-Maf, which synergize to promote Tr1 differentiation and IL-10 production[257,258]. While several transcription factors seem important for Tr1 differentiation, their master transcription factor remains to be identified.

Th3 cells were initially discovered in the context of oral tolerance[259]. They mediate their main suppressive effects via the production of TGF-β and express LAP and GARP[259–261]. Th3 cells also express low levels of IL-10 in conjuction with IL-4[262]. Similarly to Tr1 cells, this cell subset is induced peripherally after antigen stimulation[260]. The limited number of phenotypical markers identified so far have made it difficult to uncover the mechanisms underlying Th3 differentiation and whether they truly represent a distinct suppressive T cell subset.

1.5. Regulatory T cell specialization

As previously described, environmental and functional cues instruct the adaptation of helper T cells, which leads to the differentiation of distinct functional helper T cell subsets. In order to control distinct responses, Treg cells act similarly by adapting their phenotype to the requirements of these responses, and differentiate into specialized Treg cell subsets (**Figure 17**).

1.5.1. Subset-specific regulatory T cells

Under certain environmental circumstances, Treg cells acquire the expression of helper T cell transcription factors in addition to Foxp3 expression. The nature of these "hybrid" Treg subsets has been questioned. The expression of pro-inflammatory transcription factors by Treg cells was initially thought to identify an unstable Treg phenotype, associated with decreased suppressive functions. It was also thought to indicate the conversion from a Treg to a conventional T cell phenotype, leading to the differentiation of Treg cells into 'ex-Tregs', endowed with the ability to drive inflammation by producing pro-inflammatory cytokines. Alternatively, and while these statements hold true in certain contexts, most of these Treg cells retain *in vitro* suppressive activity and usually produce smaller amounts of pro-inflammatory cytokines, thus opposing a pro-inflammatory role for

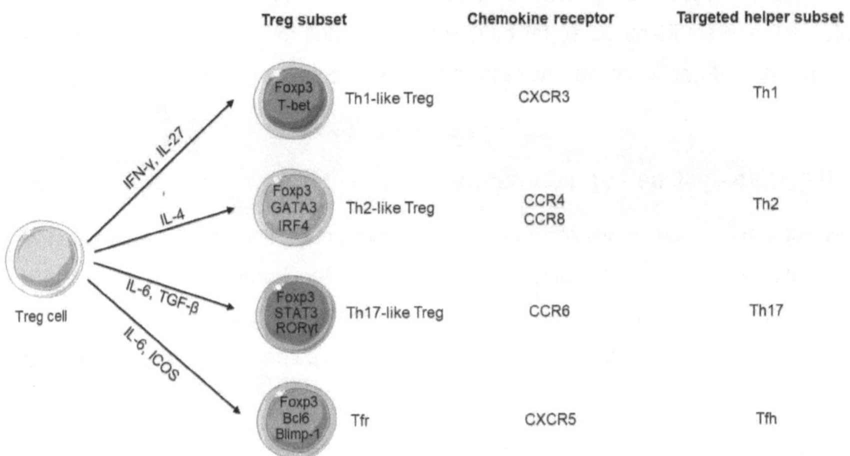

Figure 17. **Specialized regulatory T cell subsets.**
Treg cells adapt their phenotype by expressing canonical helper T cell transcription factors in order to tailor their function to their target subset. Specific subsets upregulate chemokine receptors, allowing them to co-localize with their target cells.

these cells. Accumulating evidence suggests that Treg cells express and use canonical T helper cell-associated transcription factors to regulate specific types of effector Th cells[263–270]. This confers Treg cells with the ability to traffic to the appropriate tissue, as well as co-localize and function in proximity to their target effector cells.

1.5.1.1. Th1-biased Treg cells

Although absent at steady state, T-bet-expressing Treg cells differentiate in the context of Th1 immune responses, in response to STAT1-inducing cytokines, like IFN-γ and IL-27[265,271]. T-bet directs the upregulation of chemokine receptor CXCR3 on the surface of Treg cells as well as a small increase in IFN-γ expression. CXCR3 is characteristically expressed by Th1 cells. Its expression by Treg cells allows them to respond to CXCL9, CXCL10, and CXCL11, and co-localize with Th1 cells at the sites of inflammation. The specific deletion of T-bet in Treg cells does not affect the *in vitro* suppressive activity of Treg cells[272]. T-bet deletion only leads to a modest decrease in the mRNA expression of effector molecules IL-10 and TGF-β by Treg cells. However, it prevents Treg accumulation at the sites of Th1 inflammation, thus leading to their inability to exert suppression *in situ*[265]. Indeed, fluorescent visualization has shown that T-bet⁺ CXCR3⁺ Treg cells are in closer proximity to Th1 cells and CD8 T cells than T-bet⁻ CXCR3⁻ Treg cells[273]. Similarly, Treg-specific deletion of CXCR3 in a mouse model of crescentic glomerulonephritis resulted in reduced Treg recruitment to the kidney and an overwhelming Th1 immune response[274]. This suggests that T-bet expression primarily endows Treg cells with the ability to suppress Th1 responses by controlling CXCR3 expression and their localization.

1.5.1.2. Th2-biased Treg cells

The expression of Th2-associated cytokines IL-4, IL-5, and IL-13 has been detected in Treg cells and can be associated with the development of inflammatory disorders in mice[275–278]. The expression in Treg cells of the Th2-associated transcription factor IRF4[279] is nevertheless associated with the suppression of Th2 responses. Indeed, the Treg-specific deletion of IRF4 leads to the development of an uncontrolled Th2 response, characterized by increased IL-4 and IL-5 production, and an increase in IgE and IgG levels in the serum[270]. IRF4-deficient Treg cells show a reduced expression of CCR8, a chemokine receptor required for the migration to the sites of Th2 inflammation[270]. IRF4 deficiency in Treg cells also leads to a decrease in IL-10 expression[270], possibly due to a decreased Blimp-1 expression[280], required for optimal IL-10 expression in Treg cells. Interestingly, IL-10 has been shown to be important for the regulation of Th2 responses[217]. The effects of IRF4 on Treg cells are also mediated via transcription factor JunB, which contributes to the expression of CTLA-4[281]. JunB also contributes to the expression of ICOS, a co-stimulation molecule required for the survival of effector Treg cell subsets[282].

The canonical Th2 transcription factor GATA3 can also be expressed by Treg cells[267,283,284]. GATA3 regulates many Th2-specific genes in Treg cells, such as *Il1rl1*, which encodes the IL-33 receptor, ST2, as well as CCR8[285]. GATA3+ Treg cells do not, however, express IL-13, which might be suppressed by the concomitant expression of Foxp3. Accordingly, a modest reduction in Foxp3 in Treg cells results in their production of Th2 cytokines[286]. GATA3 deficiency in Treg cells led to a decreased expression of Foxp3, as well as CTLA-4, GITR, and CD25[284]. GATA3 deficiency also led to the development of late-onset systemic immunity, with increased production of Th1, Th2, and Th17-associated cytokines[165,267,284,287]. These observations suggest that the role of GATA3 expression in Treg cells might extend beyond the control of Th2 responses.

1.5.1.3.   Th17-biased Treg cells

The requirement of TGF-β for the differentiation of both Treg and Th17 cells initially suggested a mutual exclusion of these two subsets and their respective master transcription factors, Foxp3 and RORγt[288]. Recently, Treg cells have been shown to express Th17-associated markers and transcription factors along with Foxp3, raising questions as to the phenotype of these cells.

RORγt-expressing Treg cells have been reported in a number of situations[118,268,269,288–304]. RORγt+ Treg cells can develop in the context of autoimmunity and cancer[293,296–300,302,304]. In humans, RORγt+ Treg cells have been shown to express IL-17A and, only transiently, lose suppressive function[292]. Yang et al hypothesized that RORγt+ Treg cells could represent an intermediate stage of Th17 differentiation, unstable Treg cells that can convert to IL-17-producing pathogenic cells under inflammatory conditions, or a subset of activated Treg cells generated under a Th17-prone microenvironment[304]. Using fate-mapping reporter mice, they found that RORγt+ Treg cells had a stable phenotype *in vivo*, only marginally contributed to Th17 generation during inflammation, and exhibited high expression levels of suppressive markers, suggesting that RORγt[+] Treg cells form a stable subset of Treg cells with increased immunosuppressive capacity[304].

RORγt[+] Treg cells are also found at steady state in the intestine[268,269,301,303,305–310]. Intestinal RORγt[+] Treg cells are highly suppressive, with corresponding levels of suppressive markers[268,269,303]. They are induced in the intestine in response to microbial signals and are absent in germ-free mice[268,269]. RORγt[+] Treg cells are the highest producers of IL-10 among intestinal Treg cells[289], and are thus well equipped to control intestinal Th17 responses[218,219]. RORγt-deficient Treg cells were shown to be unable to control Th1/Th17[269] and Th2[268] inflammation in the context of induced colitis. It is interesting to note that RORγt deficiency in Treg cells did not lead to the apparition of spontaneous colitis symptoms[307].

RORγt[+] Treg cells express CCR6[268,303], a chemokine receptor required for the localization of Th17 cells[311]. CCR6 has been shown to also control Treg accumulation at the sites of Th17-mediated inflammation[312]. CCR6-deficient Treg cells are unable to control colitis and EAE development[313,314]. Transcription factor STAT3 is also required for the optimal expression of CCR6, IL-35 and IL-10 by Treg cells[263]. STAT3-deficient Treg cells are unable to prevent Th17-mediated colitis[263], which might be explained by their altered IL-10 production[218,219]. STAT3 deficiency does not impact CTLA-4, which might account for the normal *in vitro* suppressive capacity of STAT3-deficient Treg cells[263]. In summary, while RORγt[+] Treg cells contribute to inflammation in certain contexts, they mainly constitute a stable and highly suppressive subset at steady state, which, by expressing CCR6 and IL-10, seems particularly adapted to suppress Th17 responses.

1.5.1.4.   Follicular Treg cells

Thymically derived Treg cells are able to mimic the Tfh phenotype[264,266]. Follicular Treg (Tfr) cells express transcription factor Bcl6, which allows them to acquire the expression of chemokine receptor CXCR5 and home to the follicular region. Similarly to Tfh cells, Tfr cells are induced in response to SLAM-associated protein, CD28, and B cell-dependent signals[264,266]. Tfr cells control the germinal center reaction by limiting Tfh and germinal B cell numbers[264,266,315]. Tfr cells express higher CTLA-4 protein levels and IL-10 mRNA levels, but show lower expression of granzyme B compared to CXCR5⁻ PD1⁻ Treg cells[266]. Tfr cells produce IL-10, but their IL-10 production is unlikely to constitute their principal mode of germinal suppression, as IL-10 is required for the stimulation of germinal center B cells[316]. Tfr cells also express Blimp-1, which is counterintuitive, given its role as a mutual repressor of Bcl6[264,266]. Tfr cells are also found in the Peyer's patches, an immune structure associated to the intestinal epithelium and involved in the production of IgA antibodies[317]. Intestinal Treg cells are peripherally induced, suggesting that Tfr cells found in the Peyer's patches could be of peripheral origin. It has been demonstrated that STAT3 is required for the differentiation of Tfr cells after immunization or in the Peyer's patches[318].

1.5.1.5.   Treg stability and plasticity

The plasticity of Treg cells endows them with incredible powers of adaptation and seems useful for the fine control of distinct immune responses. However, this plasticity seems tied to the risk of Treg instability, steering Treg cells away from their suppressive phenotype and towards the acquisition of pro-inflammatory functions. Indeed, under certain conditions, Foxp3 expression is destabilized, which compromises Treg function. The cells that have lost Foxp3 expression are termed 'ex-Tregs'.

As previously mentioned, the stability of Treg cells relies on the epigenetic status of the CNS2 within the *Foxp3* locus. Treg subsets expressing different canonical T helper transcription factors were initially thought to present a destabilized phenotype but have since then been shown to exhibit a demethylated CNS2[303,319]. In Treg cells, positive feedback loops likely maintain high levels of Foxp3 and thus contribute to the stability of the Treg phenotype, but this meta-stable state can be disrupted under strong pro-inflammatory conditions. Cytokines can destabilize Treg cells *in vitro*[320] while inflammation can shift the balance between Treg and effector T cells[321,322]. IL-6 has been shown to induce re-methylation of the CNS2 and to close the chromatin of the *Foxp3* locus[277]. Acute *Toxoplasma gondii* infection led to the loss of Foxp3 expression by Treg cells and to their acquisition of a pathogenic phenotype[323].

1.5.2. Environmental specialization

For a long time, Treg cell biology was mainly studied in the spleen and other secondary lymphoid organs. In recent years, more attention has been imparted to tissue-resident

Treg cells. Treg cells are present throughout the body and represent about 10% of the CD4 T cell pool in most tissues. It was discovered that, in addition to tailoring their function to effector T helper cell targets, Treg cells more broadly adapt their phenotype to their tissue of residence, by upregulating tissue-specific transcription factors, leading to the differentiation of tissue-specific Treg subsets.

1.5.2.1.   Tissue-specific Treg subsets

Aside from their suppressive function, Treg cells have been shown to promote tissue physiology and repair.

Treg cells control metabolic function. Depletion of Treg cells led to increased expression of proinflammatory markers in the adipose tissue and increased insulin resistance, which was rescued upon *in situ* Treg expansion[324]. Treg cells found in visceral adipose tissue form up to 50% of the CD4 T cell pool and upregulate the expression of peroxisome proliferator-activated receptor (PPARγ), a transcription factor expressed by adipocytes cells and involved in their differentiation[325,326]. The expression of PPARγ in Treg cells leads to their accumulation in the adipose tissue and the expression of lipid metabolism genes, allowing them to survive in lipotoxic environments, and enabling them to use fatty acids as metabolic fuel[325].

Treg cells also promote the proliferation and differentiation of epithelial stem cells. Skin Treg cells preferentially localize in hair follicles and promote hair follicle regeneration by augmenting hair follicle stem cell proliferation and differentiation, by expressing high levels of Jagged 1[327].

Treg cells have been shown to regulate tissue repair and regeneration in the muscle. The injured skeletal muscle and the infected lung harbor a population of amphiregulin-expressing Treg cells[328,329]. Amphiregulin is a member of the Epidermal Growth Factor family and is known to promote healing and regeneration in the muscle, the intestine, and the lung[330]. Amphiregulin deficiency in tissue Treg cells leads to severe tissue damage but does not alter Treg suppressive function[329]. Amphiregulin production is induced by the detection of cytokines IL-33 and IL-18. Both cytokines are released and activated upon tissue damage. A similar subset of Treg cells has been identified in the intestine[331].

1.5.2.2. Intestinal immunity

The intestine is a complex environment. Its principal physiological function is to take part in the digestion and the absorption of food. As a result of its function, the intestine is a major entry point for pathogens and other threats and has to be protected by a specialized immune system.

1.5.2.2.1. Intestinal architecture

The small intestine is divided into duodenum, jejunum and ileum, and is responsible for the absorption of nutrients into the bloodstream[332] (**Figure 18**). The large intestine is divided into cecum, or appendix in humans, colon, and rectum. It harbors most of the gut microbiota and is necessary for water absorption and the excretion of undigested foods.

The intestinal tract is divided into distinct regions: the lumen, where most of the gut microbes resides; the mucosa, siege of most intestinal immune responses; the submucosa, a dense layer of connective tissue that contains numerous parasympathetic nerves; the muscularis propria, a layer of smooth muscle involved in intestinal contractions; and the adventitia or serosa, a thin fibrous covering that segregates the intestine within the peritoneum (**Figure 19**). The mucosa consists of an epithelial layer, the lamina propria, and the muscularis mucosa, a thin muscle layer. The intestinal lamina propria (LP) is composed of connective tissue, blood, and lymphatic vessels, nerve endings, and harbors most of the immune cells. The intestine is also associated with several gut-associated lymphoid tissues (GALT). Mesenteric lymph nodes form a chain of lymph nodes which drain different parts of the intestine. Peyer's patches in the small intestine, cecal patches, colonic patches, and isolated lymphoid follicles are intestinal immune structures dedicated to sampling and inducing adaptive immune responses. Peyer's patches are notably involved in the production of IgA antibodies[333].

The intestinal epithelium is lined with tightly joined epithelial cells and shows structural adaptations aimed at increasing its absorptive surface[334]. Plicae are large circular fold of the mucosa, most numerous in the upper part of the small intestine. Villi are finger like mucosal projections forming smaller folds and microvilli are fine projections found at the apical surface of enterocytes. The intestinal epithelium also harbors specialized cells.

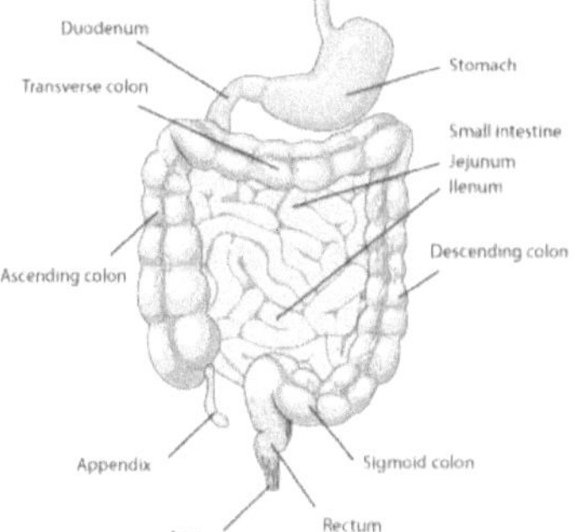

Figure 18. Architecture of the intestine.
The small intestine serves for digestion and nutrient absorption and has three sub-regions: the duodenum, the jejunum and the ileum. The large intestine helps with water absorption, excretion of food waste and harbors most of the gut microbiota. It is composed of the cecum, in mice, or the appendix in humans, the colon, and the rectum.

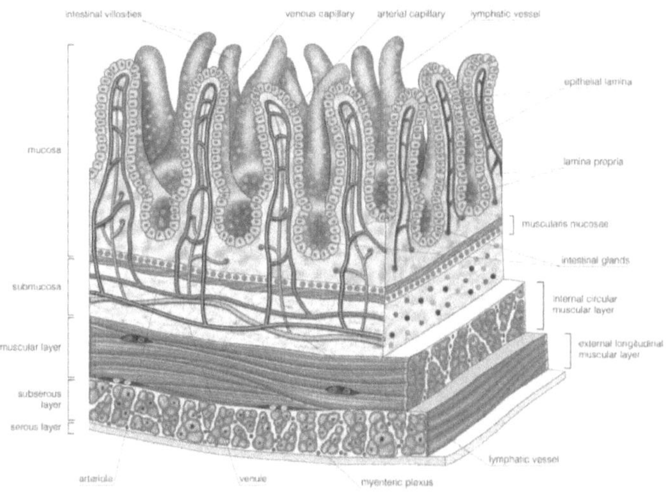

Figure 19. Histological structure of the intestinal barrier.
The intestinal tract is divided into the lumen, the mucosa, the submucosa, the muscularis propria, and the serosa. The mucosa, where most immune cells are localized, consists of an epithelial layer, the lamina propria, and the muscularis mucosa.

Paneth cells are specialized secretory epithelial cells, located in the crypts of the small intestine[335]. They secrete antimicrobial peptides, basic amino acid-rich cationic small proteins, which protect the host against infection by Gram-positive and Gram-negative bacteria. Goblet cells produce gel-forming mucins. These heavily glycosylated proteins interact to form mucus, which physically prevents microbial invasion of the mucosa[336]. Enteroendocrine cells are stimulated by products of food digestion and microbial fermentation to produce gut hormones that coordinate food digestion and absorption, insulin secretion and appetite[337]. Tuft cells are rare cells found in the small intestine. They act as chemosensory sentinels and have recently been shown to interact with the immune system[338]. A major component of the intestinal epithelium are intraepithelial lymphocytes (IELs). They consist of various T cell subsets that interact with enterocytes in order to maintain normal homeostasis[339]. The epithelium of the large intestine differs from the small intestine. It is covered with two mucus layers, protecting it from the increased microbial concentrations[336]. While the inner mucus layer is tightly structured and does not allow the penetration of gut bacteria, the outer mucus layer is looser and harbors numerous bacterial species, which use mucin as an energy source.

1.5.2.2.2.   Gut microbiota

The intestine harbors up to $3.8.10^{13}$ bacteria, mainly concentrated in the large intestine and distributed in a gradient from the small intestine to the rectum[340]. The gut microbiota plays a crucial role in digestion and metabolism, by fermenting complex carbohydrates and synthetizing vitamins and metabolites[341,342]. The microbiota and its metabolites also play a role in the maintenance of gut epithelial integrity[343]. The microbiota shapes the mucosal immune system, and helps balance defense against microbial pathogens and oral tolerance to the microbiota[3,343,344]. The microbiota impedes the colonization of pathogens by competing for nutrient sources and physical space, but some species also produce inhibitory metabolites or directly kill competitors, thus preventing the establishment of harmful pathogens[13].

1.5.2.2.3.   Intestinal immune system

The intestinal immune system is tasked with the defense against pathogens but also plays a major role in maintaining tolerance to commensal symbionts and dietary antigens

(**Figure 20**). The intestinal environment can be categorized into inductive and effector sites. Accordingly, intestinal immune responses are induced in the GALT and then migrate to the lamina propria via the lymphatic system to exert their immune functions.

Due to the high concentration of microbial antigens, the intestinal environment is characterized by a type 3 immune response[345,346] at steady state, and is thus mainly mediated by RORγt[+] cells. DCs sample antigens through the intestinal barrier and migrate to GALT, where they activate helper T cells. Helper T cells subsequently upregulate the mucosal homing receptors α4β7 and CCR9 and migrate to the lamina propria[347]. Th17 cells form the most abundant helper T cell subset in the lamina propria. They are tasked with controlling microbiota composition and eliminating harmful microbial species[348]. Germ free mice exhibit impaired immune responses, indicating that the microbiota has a direct effect on intestinal immune system development[349]. Indeed, the microbiota promotes the development and priming of neutrophils, macrophages, helper and regulatory T cells, innate lymphoid cells and IgA production[13]. Innate lymphoid cells and γδ T cells also play a role in maintaining epithelial integrity and fighting off pathogens, by producing antimicrobial peptides, and cytokines IL-17A and IL-22[346]. Intestinal DCs also promote the differentiation of intestinal Treg cells by producing TGF-β and retinoic acid[350]. Intestinal Treg cells play a major role in maintaining intestinal immune homeostasis and can be divided into several specialized subsets based on their phenotypical markers and function.

1.5.2.3. Intestinal Treg cells

The proportion of Treg cells found in the intestine is higher than that of other organs. Indeed, while it averages about 10% in other organs, Treg cells constitute around 20% of the total CD4 T cell population in the small intestine lamina propria (siLP), and more than 30% in the colon lamina propria (cLP)[334].

Intestinal immune cells have to face three major threats: dietary antigens, tissue damage, and microbial antigens. Accordingly, Treg cells have adapted their phenotype in order to suppress the specific immune responses elicited by these different threats. Three intestinal Treg subsets have been identified based on transcriptomic and functional data, and can be distinguished by their expression of transcription factors GATA3 and RORγt,

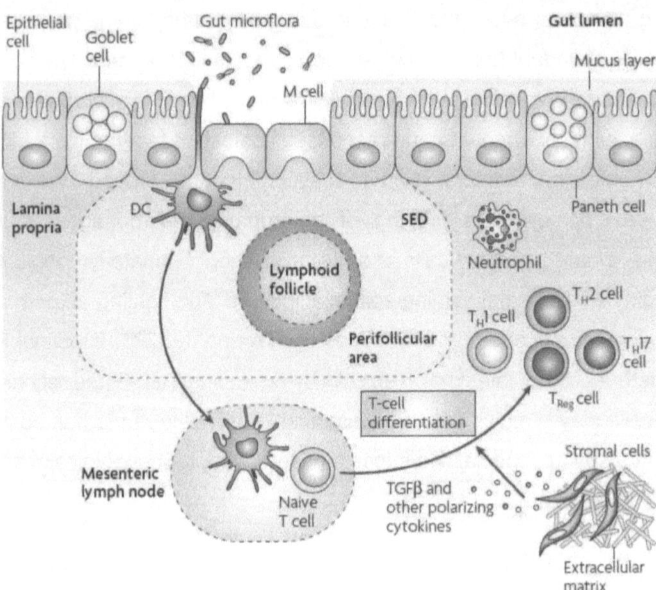

Figure 20. **Intestinal immune system.**
The intestinal immune system is tasked with the defense against pathogens but also plays a major role in maintaining tolerance to commensal symbionts and dietary antigens.
DCs sample intestinal luminal contents and are able to activate naïve T cells in GALT, which will mainly differentiate into anti-bacterial Th17 cells, but can also differentiate into Treg cells, or Th1 and Th2 cells. SED, subepithelial dome; TGFβ, transforming growth factor-β; TH, T helper; TReg, T regulatory.
From: Cho, Judy H. "The Genetics and Immunopathogenesis of Inflammatory Bowel Disease." *Nature Reviews Immunology* 8, no. 6 (June 2008): 458–66.

the master regulators of Th2 and Th17 cells, respectively. These subsets are GATA3[+] Treg cells, RORγt[+] Treg cells, and GATA3[-] RORγt[-] Treg cells (**Figure 21**).

a. GATA3[+] Treg cells

GATA3[+] Treg cells form about a third of Treg cells in the small intestine and in the colon. GATA3 expression is conserved in germ-free and antigen-free mice, suggesting a thymic origin of this population[267,331,351]. This is confirmed by the high expression levels of Nrp1 and Helios among this subset[269]. Intestinal GATA3[+] Treg cells express high levels of the activation markers KLRG1, CD103, and OX40[331]. They also express the IL-33 receptor, ST2, which is regulated by GATA3 in a feed-forward manner[331]. ST2[+] Treg cells show enhanced activation profiles and production of cytokines IL-10 and TGF-β[352]. IL-33, secreted by IELs and enterocytes upon tissue damage, activates ST2[+] Treg cells, allowing them to restrain responses to auto-antigens upon injury[353,354]. Similarly to Treg cells found in the muscle or the lung, intestinal GATA3[+] Treg cells express high levels of amphiregulin, a protein involved in tissue repair[331,355].

b. RORγt[+] Treg cells

RORγt[+] Treg cells are mainly found in the colon, where they constitute up to 50% of the Treg pool, and up to 30% of cells in the small intestine[334]. These cells express low levels of Nrp1[303] and Helios[268,269], suggestive of their peripheral origin. Regardless, RORγt[+] Treg cells exhibit a highly demethylated CNS2, indicative of a stable Treg phenotype[303]. RORγt[+] Treg cells express increased levels of ICOS, CTLA-4, ectonucleotidases CD39 and CD73, and are the highest-producing subset of IL-10 among intestinal Treg cells[268,269], indicating a highly suppressive phenotype. Unlike Th17 cells, intestinal RORγt[+] Treg cells do not express IL-17A at the protein level, even though *Il17a* transcripts are detectable[303,307]. Two studies have involved RORγt[+] Treg cells in the regulation of either Th2[268] and Th1/Th17[269] intestinal responses. The diverging results of these independent studies can be attributed to variations in experimental protocols, but also animal housing and microbiota compositions. However, as previously mentioned, RORγt[+] Treg cells colocalize with Th17 cells and express high levels of IL-10, suggesting an adaptation to the suppression of antimicrobial immune responses.

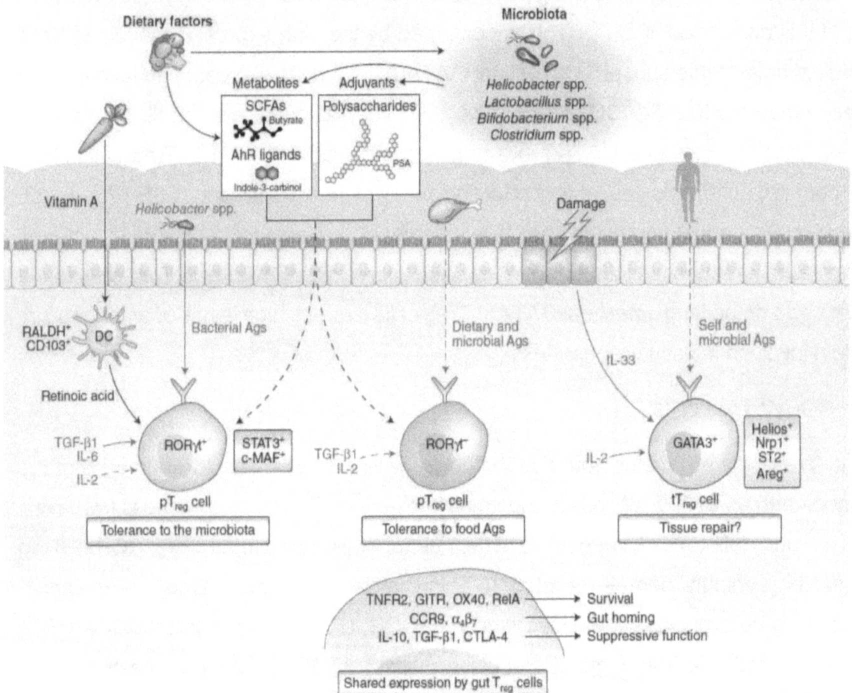

Figure 21. **Intestinal Treg subsets.**
Intestinal Treg cells are highly specialized and can be divided into three subsets. RORγt⁺ Treg cells are induced by the microbiota and mediate tolerance to the microbiota. RORγt⁻ Treg cells seem to depend on dietary antigens for their induction and mediate tolerance to food antigens. GATA3⁺ Treg cells are tTreg cells and are induced in response to IL-33, produced in response to tissue damage. Dashed arrows indicate that further evidence is needed for conclusive results; red arrows indicate the TCR specificity of the Treg cell subsets.
From: Whibley, Natasha, Andrea Tucci, and Fiona Powrie. "Regulatory T Cell Adaptation in the Intestine and Skin." *Nature Immunology* 20, no. 4 (April 2019): 386–96.

RORγt[+] Treg cells are heavily reduced in germ-free mice and after treatment with broad-spectrum antibiotics[356,357], indicating that they are induced by microbial antigens. This subset can be restored upon microbiota recolonization[356,357]. However, the ability of microbial species to induce RORγt[+] Treg cells varies[269] and depends partly on their SCFA production and their ability to inhibit histone deacetylases, which increases Foxp3 expression. Of note, inflammatory bowel disease (IBD) patients show a decrease in colonic butyrate-producing bacteria and mucosal butyrate transporter levels[358,359], suggesting an alteration of RORγt[+] Treg induction. Other environmental factors have been involved in the induction of RORγt[+] Treg cells, such as cytokines TGF-β, IL-6, and IL-23, and retinoic acid[268,269,305]. The induction of RORγt[+] Treg cells has very recently been shown to act in an antigen-specific manner. Indeed, while colonic T cells with TCRs recognizing SFB led to the differentiation of Th17 cells, colonic T cells recognizing the pathobiont *Helicobacter hepaticus* differentiate into RORγt[+] Treg cells under homeostatic conditions[307].

c. GATA3[-] RORγt[-] Treg cells

GATA3[-] RORγt[-] Treg cells control immune responses directed towards food antigens. This subset is mainly found in the small intestine, as it is the seat of nutrient absorption. It constitutes about 50% of siLP Treg cells against 15% of cLP Treg cells. GATA3[-] RORγt[-] Treg cells are induced in response to dietary antigens and not microbial antigens. Indeed, germ free mice do not exhibit decreased numbers of GATA3[-] RORγt[-] Treg cells, contrary to antigen-free mice[351]. Most of these cells express low levels of Nrp1 and Helios, suggesting a peripheral origin. Double negative intestinal Treg cells are involved in oral tolerance and seem to control food allergy. However, further transcriptomic and phenotypic analyses are needed to better define the functional reach of this cell subset as well as identify characteristic phenotypical markers.

While GATA3[+] Treg cells have been observed in human blood[267], and RORγt[+] Treg cells have been identified in the colon of healthy controls and Crohn's disease patients[269], the phenotype of human intestinal Treg cells and whether it mirrors murine Treg cell populations is still unclear. The functional and phenotypical specificities of intestinal Treg cell subsets, both in human and mouse, as well as the molecular mechanisms responsible for keeping the balance between the different specialized subsets, are yet to be fully elucidated.

1.6. Transcription factor c-Maf

Recently, transcription factor c-Maf has been identified as a major transcription factor in the control of Tfr and RORγt[+] Treg cell differentiation and function[305,307].

The role of transcription factor c-Maf in the differentiation, function, and homeostasis of T cells has been recently reviewed by us[308] (**Annex 1**). This next section will briefly summarize some essential aspects of the structure and function of c-Maf.

1.6.1. General concepts

The *Maf* gene encodes transcription factor c-Maf (**Figure 22**). The v-Maf oncogene was initially identified in chickens developing musculoaponeurotic fibrosarcoma after infection with the replication-defective retrovirus AS42[360–362]. Its cellular counterpart, c-Maf, has since then been involved in multiple developmental and physiological processes[363–374], which accounts for the lethality of the full c-Maf KO in mice[363,365,371]. Indeed, the physiological roles of c-Maf extend to numerous organs and are notably involved in lens[363–365], neural[366,374], renal[367], bone[368–370], blood[371], and pancreas development[372,373].

c-Maf is a basic leucine zipper belonging to the AP-1 superfamily of transcription factors. c-Maf binds to other partners via its leucine zipper domain and forms homo- or heterodimers, with compatible proteins such as Jun and Fos[375–377]. c-Maf can also bind to other non-bZIP proteins, such as Sox proteins[368]. The basic region of dimeric Maf factors allows them to recognize a palindromic sequence referred to as the Maf Recognition Element (MARE)[378].

In T cells, TCR stimulation induces c-Maf transcription. The maintenance of c-Maf expression requires additional stimuli, in the form of co-stimulatory signals or cytokines. TGF-β and IL-6 induce c-Maf in Th17 cells[307,379–381], which exhibit the highest c-Maf levels among T cells, in a STAT3-dependent manner. IL-4 induction of c-Maf in Th2 cells also depends on STAT3[382,383], and not on STAT6, as was formerly believed. ICOS promotes

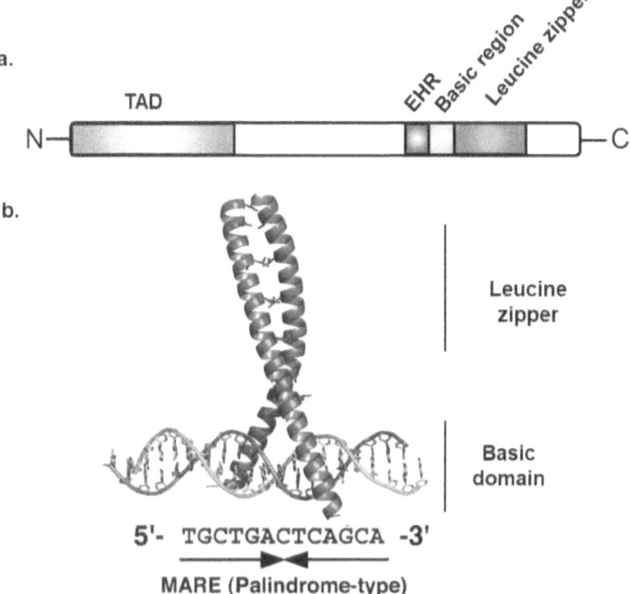

Figure 22. **Transcription factor c-Maf.**
a. Transcription factor c-Maf contains four distinct functional domains: an N-terminal transactivation domain (TAD), a basic region and an extended homology region (EHR), and a C-terminal leucine zipper domain. b. c-Maf binds to DNA at MARE sites with its basic domain and forms dimers with its leucine zipper domain.
Adapted from: From: Imbratta, Claire, Hind Hussein, Fabienne Andris, and Grégory Verdeil. "C-MAF, a Swiss Army Knife for Tolerance in Lymphocytes." *Frontiers in Immunology* 11 (2020).

the expression of c-Maf, in Th2, Th17, and Tfh cells[127,384–386]. IL-27 signals both via STAT1 and STAT3 and induces expression of c-Maf during Tr1 cell differentiation[257,258].

c-Maf expression is post-transcriptionally regulated by several miRNAs[387–390], but these miRNAs have not yet been reported in T cells. c-Maf is also the object of several post-translational modifications. Phosphorylation of c-Maf is required for nuclear translocation and binding to the promoter of target genes[391–395], while SUMOylation of c-Maf is negatively associated with its transactivating potential[396,397].

1.6.2. Role of c-Maf in effector T cells

c-Maf was formerly introduced as a Th2 transcription factor, however its roles now extend to most known T cell subsets (**Figure 23**), lending it a crucial position in adaptive immunity and anti-tumoral responsiveness.

Pioneer studies identified c-Maf as a regulator of IL-4 expression in Th2 cells, in collaboration with GATA3, STAT6, and NFAT[365,398–400]. Regardless of its role in IL-4 induction, c-Maf has an inhibitory role over both Th1 and Th2 inflammatory responses[401–403]. c-Maf has also been shown to repress IL-2 expression[401], which leads to increased Th17 differentiation. In naïve T cells, c-Maf associates with Sox5 and directly induces the expression of *Rorc*[404], leading to the differentiation of Th17 cells. c-Maf globally acts as a negative regulator of pro-inflammatory Th17 genes[405], but is involved in the induction of IL-21[406], the maintenance of Th17 cells and their stabilization through the expression of IL-23R[127]. c-Maf is essential for the development of Tfh cells[129,406] and could contribute to the induction of Bcl6 in Tfh cells[406]. c-Maf is also involved in the adequate IL-4 and IL-21 expression in Tfh cells[129,407,408], two key cytokines in Tfh function.

Besides its subset specific roles, c-Maf has a common regulatory function over IL-10 expression in several T cell subsets. c-Maf is a direct positive regulator of *Il10* in distinct helper T cell subsets[401]. In order to optimally induce IL-10 expression, c-Maf interacts with different transcription factors, like AhR in Tr1 cells[258], or Blimp-1 in Th1 cells[409]. c-Maf also provides a common mechanism for a negative regulation of IL-2 signaling *in vivo* in models of Th1, Th2, and Th17 responses[401].

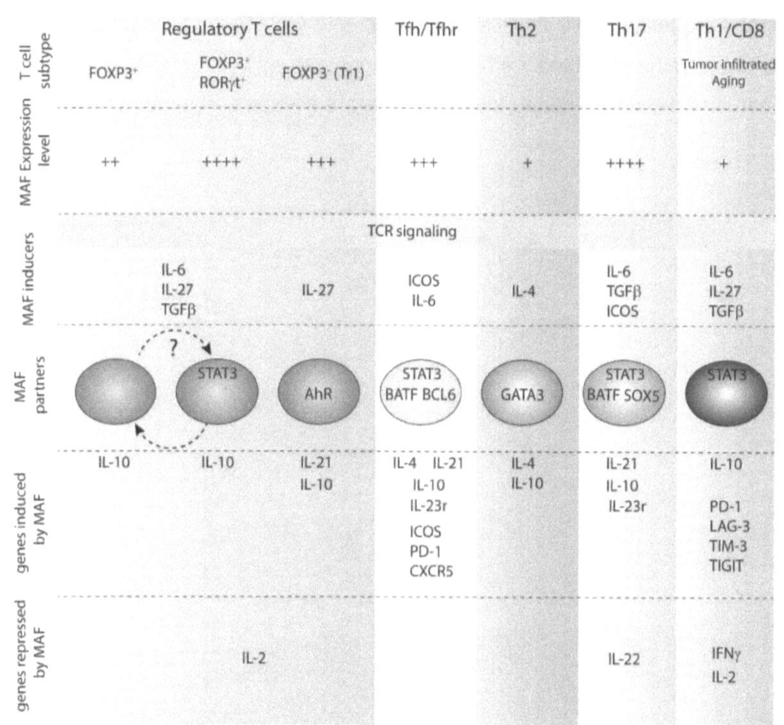

Figure 23. **Role of c-Maf in various T cell subets.**
This figure shows the level of c-Maf expression, in addition to its induction signals, its protein partners, and its target genes in distinct T cell subsets.
From: Imbratta, Claire, Hind Hussein, Fabienne Andris, and Grégory Verdeil. "C-MAF, a Swiss Army Knife for Tolerance in Lymphocytes." *Frontiers in Immunology* 11 (2020).

CD8 T cells do not express c-Maf at steady state. c-Maf expression is however detected in tumor-infiltrating lymphocytes (TIL)[410]. Overexpression of c-Maf in CD8 T cells leads to a strong repression of IFN-γ and IL-2 production, and to an increased expression of genes associated with T cell exhaustion, a dysfunctional state observed during chronic infections or in TILs[411,412]. c-Maf cooperates with Blimp-1 to induce the expression of inhibitory receptors in CD8 T cells[413].

All in all, c-Maf expression in T cells is associated with increased tolerogenic functions, which contributes to the maintenance of immune homeostasis in inflammatory conditions, but which can also be used by tumors to favor immune escape and tumor development.

Chapter 2: Research objectives

In recent years, the study of tissue-resident Treg cells has revealed the existence of multiple functional and tissular Treg specialization programs, amounting to a much more heterogenous view of the Treg phenotype than previously imagined. This is highlighted in the intestine, where the complex intestinal environment has led to the adaptation and cohabitation of three distinct intestinal Treg subsets. This questions which molecular mechanisms govern the regulation of different Treg subsets present in the same tissue and how this is shaped by inflammatory processes.

We previously reported a role for transcription factor c-Maf in the differentiation of Tfh cells in response to vaccination and in response to the intestinal microbiota[406] (**Annex 2**). In the course of this work, we observed that c-Maf was widely expressed by intestinal Treg cells, where c-Maf$^+$ cells constitute up to 90% of Treg cells. This prompted us to investigate the role of transcription factor c-Maf in Treg cell differentiation and function, with a particular focus on intestinal Treg cells.

In the first part of our work, using a conditional knockout mouse model deficient for c-Maf specifically in Treg cells (*c-Maf*$^{fl/fl}$ x *Foxp3*^{CreYFP}), we studied the role of c-Maf in Treg cell differentiation and function at steady state, particularly in the intestine. We also studied the role of c-Maf in Treg cells in the context of intestinal inflammation models and cancer models.

In the second part of our work, we studied the molecular mechanisms involved in the expression of RORγt in intestinal Treg cells, using both *in vitro* and *ex vivo* approaches.

Chapter 3: Results and discussion

Part 1: Role of transcription factor c-Maf in Treg differentiation and function

Preliminary remark

During the course of our work, several teams published their work on the role of c-Maf in Treg differentiation[305,307–309], particularly its role in the acquisition of RORγt expression and the control of Th17 intestinal immunity. These results will be discussed complementarily to our own, in order to provide a more thorough understanding of the mechanisms regulating intestinal Treg specialization, and particularly the role of c-Maf in that process.

3.1.1. c-Maf identifies a subset of effector Treg cells with an intestinal phenotype

Using flow cytometry, we analyzed the distribution of c-Maf[+] Treg cells across different lymphoid and non-lymphoid organs in naïve WT mice (**Figure 24A**). The thymus harbors the lowest frequency of c-Maf[+] Treg cells, with less than 5% of thymic Treg cells expressing this transcription factor (**Figure 24B**). In the spleen, axillary, and mesenteric lymph nodes (mLN), 30 to 40% of Treg cells express c-Maf. In non-immune organs like the lung and the liver, 20 to 30% of Treg cells express c-Maf. Interestingly, up to 90% of Treg cells residing in the small intestine lamina propria (siLP) or the colon lamina propria (cLP) express c-Maf. Our results also showed that, in comparison to conventional T cells found in other organs, intestinal conventional T cells expressed higher levels of c-Maf (**Figure 24A**). This observation could be due to the increased proportions of Th17 cells found in the intestine compared to other organs. These results indicate that c-Maf is preferentially expressed by intestinal T cells, and most particularly by intestinal Treg cells.

To characterize c-Maf[+] Treg cells, we looked at the expression of activation markers by flow cytometry. In mesenteric lymph nodes, c-Maf-expressing Treg cells do not express CD62L (**Figures 25A, 25B**), a selectin found at the surface of naïve T cells, but express high levels of activation markers CD44 (**Figures 25C, 2D**) and ICOS (**Figures 25E, 2F**).

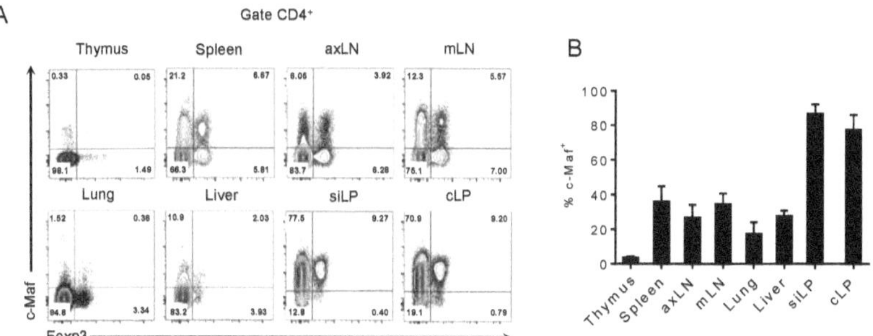

Figure 24. **c-Maf is preferentially expressed in intestinal Treg cells**. (A) Representative flow cytometry expression profiles of Foxp3 versus c-Maf of CD4 T cells in the indicated organs of naïve WT mice. (B) Histograms show the frequency of c-Maf+ cells among Treg cells in the indicated organs. Results are representative of at least three independent experiments; histograms represent the mean ± SD of at least three individual mice. (axLN: axillary lymph nodes, mLN: mesenteric lymph nodes, siLP: small intestine lamina propria, cLP: colon lamina propria)

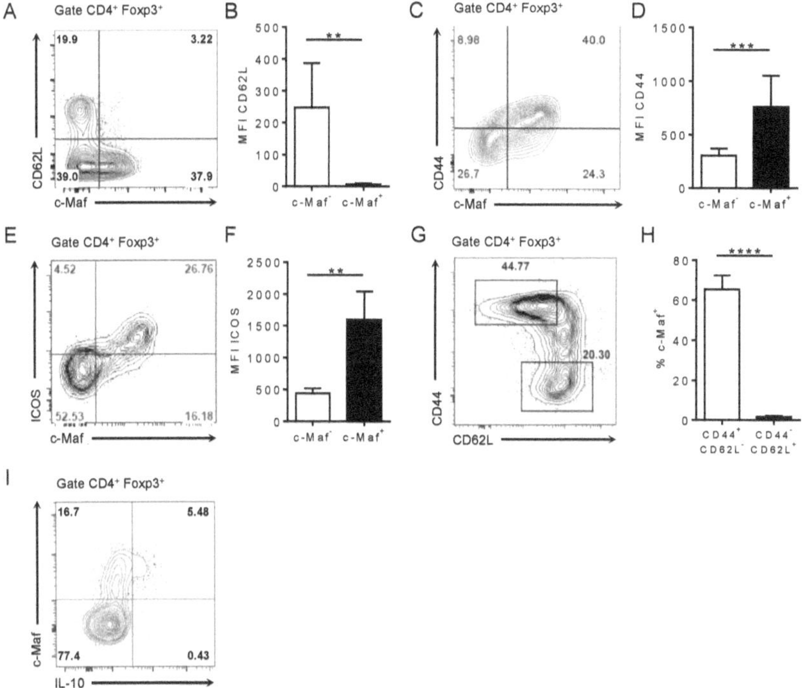

Figure 25. **c-Maf+ Treg cells have an activated phenotype**. (A - F) Expression of activation markers CD62L (A, B), CD44 (C, D), or ICOS (E, F) versus c-Maf among Treg cells in mLN of WT mice. (A, C, E) Representative flow cytometry expression profiles. (B, D, F) Histograms show the MFI of activation markers among c-Maf- and c-Maf+ Treg cells. (G, H) Expression of CD44 and CD62L among Treg cells (G) and frequency of c-Maf+ cells among eTreg (CD44+ CD62L-) and cTreg (CD44- CD62L+) cells (H) in spleen of WT mice. (I) Expression of c-Maf versus IL-10 in Treg cells in mLN of WT mice. Histograms represent the mean ± SD of at least five individual mice. Difference between groups is determined by an unpaired t test (B, D, H) or a Mann–Whitney test for two-tailed data (F). **p < 0.01; ***p < 0.001; ****p < 0.0001

CD62L and CD44 expression are routinely used to assess T cell activation status[44]. Naïve T cells are identified as CD62L⁺ CD44⁻ cells. Effector T cells lose CD62L expression and acquire CD44 and are therefore defined as CD62L⁻ CD44⁺ cells. This distinction is also applicable among Treg cells[414]. ICOS is a co-stimulatory molecule expressed mainly on CD4 T cells after activation. Accordingly, ICOS expression has been reported in a subset of effector Treg cells, characterized by a high expression of IL-10[415,416]. ICOS signaling has also been involved in the induction of c-Maf expression in Tfh and Th17 cells[127]. c-Maf⁺ cells are mainly found among the CD62L⁻ CD44⁺ effector Treg cell subset (**Figures 25G, 25H**). c-Maf notoriously promotes IL-10 expression in different conventional T cell subsets[401]. Treg cells rely on the expression of IL-10 for their suppressive function, particularly in the intestine and at mucosal surfaces[217]. Interestingly, in mLN, IL-10-expressing Treg cells were found uniquely among c-Maf-expressing Treg cells (**Figure 25I**). CD25, the high affinity chain of the IL-2 receptor, is highly expressed by Treg cells[151]. We found that c-Maf expression inversely correlated with CD25 expression in total Treg cells in the spleen (**Figures 26A, 26D**). CD25 is involved in the homeostasis of central Treg cells, unlike effector Treg cells which do not rely on IL-2 for their survival, but instead depend on continuous ICOS signaling[417]. As expected from their dependence on IL-2, most central Treg cells express CD25 (**Figure 26B**), but they do not however express c-Maf. We also observed an inverse correlation between c-Maf and CD25 expression among the effector (CD44⁺ CD62L⁻) Treg subset (**Figures 26C, 26E**). This supports the existence of a subset of c-Maf⁺ IL-10-producing effector Treg cells, which relies on ICOS signaling and not on IL-2 for its survival.

We then wanted to determine the developmental origin of c-Maf⁺ Treg cells. For that, we looked at the expression of Nrp1 at the surface of Treg cells by flow cytometry (**Figure 27**). Neuropilin 1 (Nrp1) is a surface marker which has been used to distinguish between thymic (tTreg) and peripheral (pTreg) Treg cells[202,203]. We used RORγt⁺ Treg cells as a pTreg control. Indeed, these cells are induced in the intestine by the microbiota, and accordingly express low levels of Nrp1 in mLN and siLP (**Figure 27A**), indicative of their peripheral origin. In the spleen, the great majority of Treg cells, including c-Maf⁺ Treg cells, express Nrp1, suggesting a thymic origin (**Figure 27B**). In mLN, c-Maf expression seems inversely correlated with that of Nrp1. Flow cytometry profiles nonetheless delineate two

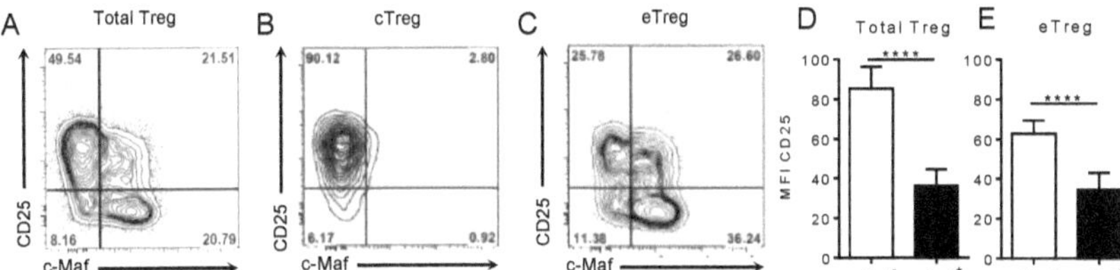

Figure 26. **c-Maf expression in Treg cells is inversely correlated with CD25**. (A – C) Representative flow cytometry expression profiles of CD25 versus c-Maf in total Treg (A; gate Foxp3+), cTreg (B; gate CD44- CD62L+ Foxp3+), or eTreg (C; gate CD44+ CD62L- Foxp3+) cells in spleen of WT mice. (D, E) Histograms show the MFI of CD25 in c-Maf- and c-Maf+ total Treg (D), or eTreg (E) in spleen of WT mice. Histograms represent the mean ± SD of nine individual mice. Difference between groups is determined by an unpaired t test. ****p < 0.0001.

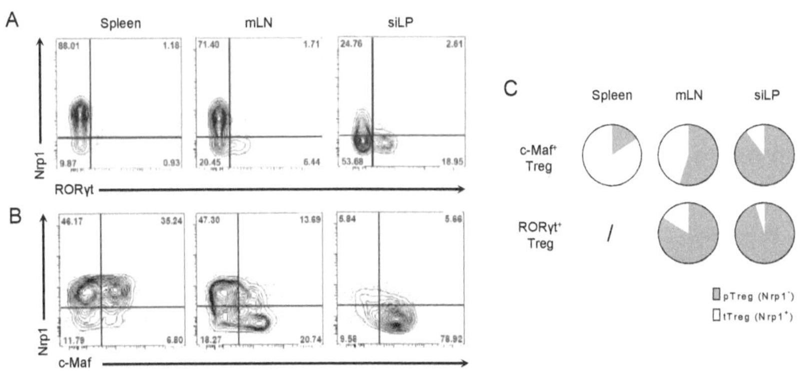

Figure 27. **c-Maf-expressing Treg cells can be of thymic or peripheral origin**. (A, B) Representative flow cytometry expression profiles of Nrp1 versus RORγt (A) or c-Maf (B) of Treg cells in the indicated organs of naïve WT mice (gate CD4+ Foxp3+ cells). (C) Pie charts show the relative frequencies of thymic (tTreg) and peripheral (pTreg) Treg cells among c-Maf+ and RORγt+ Treg subsets in the spleen, mLN, and siLP. Results are representative of at least three independent experiments. (mLN: mesenteric lymph nodes, siLP: small intestine lamina propria)

populations of c-Maf⁺ Treg cells. In the mLN, c-Maf⁺ Nrp1⁺ cells make up 15% of Treg cells, whereas c-Maf⁺ Nrp1⁻ cells constitute 20% of Treg cells. In the siLP, Treg cells express low levels of Nrp1, in accordance with their peripheral origin. Contrary to c-Maf⁺ Treg cells in the spleen, c-Maf⁺ Treg cells in the intestinal environment seem to be of peripheral origin. These results show that, contrary to RORγt⁺ Treg cells, c-Maf-expressing Treg cells can be of peripheral or thymic origin, depending on their organ of residence (**Figure 27C**). This suggests that the acquisition of c-Maf expression is not tied to the developmental origin of Treg cells.

The intestine is a complex environment which harbors distinct subsets of Treg cells. Most notably, a major subset of intestinal Treg cells expresses RORγt and controls immune responses directed towards the microbiota. The preferential expression of c-Maf among intestinal Treg cells prompted us to investigate the relative expression of c-Maf within different intestinal Treg cell subsets (**Figure 28**). While conventional T cells and RORγt⁻ Treg cells express low levels of c-Maf in mLN, RORγt⁺ Treg cells express high levels of c-Maf. Increased levels of c-Maf were observed in Tconv and Treg cells in siLP and cLP. Regardless of the organ of residence, T cells conserve the same hierarchy of c-Maf expression, with Tconv cells expressing the lowest levels of c-Maf, RORγt⁻ Treg cells expressing intermediate levels, and RORγt⁺ Treg cells unvariably expressing the highest levels of c-Maf.

These observations collectively indicate that c-Maf expression identifies a subset of effector Treg cells, preferentially found in the intestine, and characterized by the expression of RORγt, ICOS, and IL-10.

3.1.2. c-Maf is dispensable for the maintenance of central tolerance by Treg cells

Because c-Maf plays a role in a multitude of developmental processes, the complete deletion of c-Maf in mice often causes pre- or post-birth lethality, or can lead to serious developmental problems[363,365,371]. This issue is resolved by the use of conditional knockout mouse models, which additionally allow the identification of cell- or tissue-specific roles of a given gene of interest. In order to determine the role of c-Maf in the

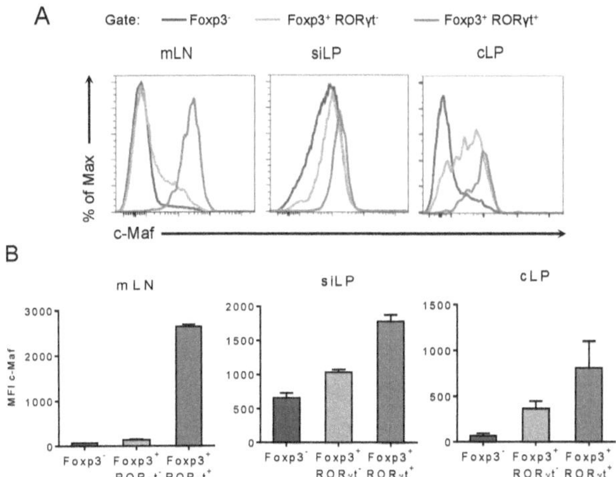

Figure 28. **c-Maf is preferentially expressed by RORγt⁺ Treg cells**. (A, B) Expression profile (A) and median of fluorescence intensity (B) of c-Maf among Foxp3⁻, Foxp3⁺ RORγt⁻, and Foxp3⁺ RORγt⁺ CD4 T cells in the indicated organs of naïve WT mice. Results are representative of at least three independent experiments; histograms represent the mean ± SD of three individual mice. (mLN: mesenteric lymph nodes, siLP: small intestine lamina propria, cLP: colon lamina propria)

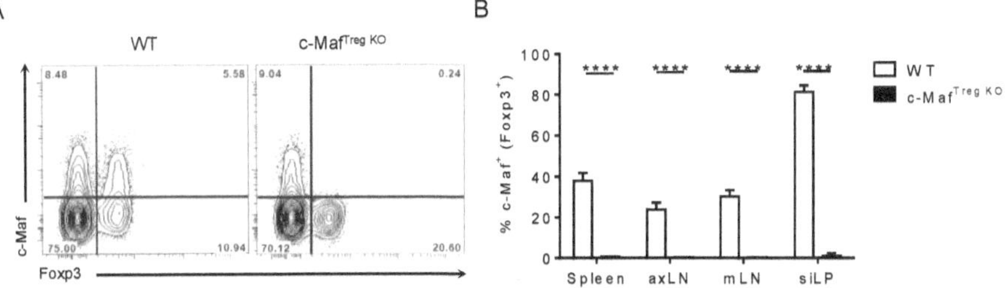

Figure 29. **c-Maf is selectively lost in Treg cells of c-Maf$^{Treg\ KO}$ mice.** (A) Representative flow cytometry expression profiles of Foxp3 versus c-Maf in CD4 cells in the mLN of WT and c-Maf$^{Treg\ KO}$ mice. (B) Histograms show the frequency of c-Maf⁺ cells in Treg cells in the indicated organs of WT and c-Maf$^{Treg\ KO}$ mice. Histograms represent the mean ± SD of five individual mice. Difference between groups is determined by a two-way ANOVA. **** p < 0.0001

differentiation and the function of Treg cells, we developed a conditional knockout mouse model by crossing c-Maf floxed mice (*c-Maf^{fl/fl}*) with mice expressing the Cre recombinase under the control of the *Foxp3* promoter (*Foxp3^{CreYFP}*). These mice, which will be further referred to as c-Maf^{Treg KO} mice, are deficient for c-Maf specifically in Treg cells.

Studies have shown that *Foxp3^{CreYFP}* allele is 'leaky', and that it can show some activity outside of the Treg lineage[418]. The expression of a fluorescent reporter gene (YFP), fused with the Cre, allows direct detection of Cre expression. However, while reporter mice are useful to determine Cre activity, Cre expression is rarely equal to Cre-mediated recombination[419]. Indeed, each locus has a distinct sensitivity to Cre recombination[419], with some loci being targeted with high efficiency, while other loci might be resistant to targeting. In extreme cases, ectopic Cre recombination can lead to germline deletion of the targeted allele[420]. Potential germline recombination could not be picked up by classical genotyping, but would require targeting the recombined allele in addition to the wild-type and floxed alleles[420].

In addition to the 'leaky' phenotype of *Foxp3^{CreYFP}* mice, the *Foxp3^{CreYFP}* allele is hypomorphic[418], which results in a decreased expression of Foxp3 in Treg cells and can directly alter Treg function. It is therefore important to use *Foxp3^{CreYFP}* as a WT control, in order to correctly interpret the experimental results. We verified the deletion of c-Maf in Treg cells by comparing c-Maf expression in Foxp3$^+$ and Foxp3$^-$ T cells between WT and c-Maf^{Treg KO} mice by flow cytometry. As shown in **Figure 29**, c-Maf expression is lost in Foxp3-expressing Treg cells and seems unaffected in Foxp3$^-$ conventional T cells. This data indicates that c-Maf is effectively knocked-out in Treg cells but not in conventional Th cells. Although it is not sufficient to formally exclude off-target recombination, we nevertheless decided to pursue our study with this mouse strain.

Mice with an altered Treg function, notably due to mutations in *Foxp3*, develop systemic autoimmunity[421]. c-Maf^{Treg KO} mice develop and reproduce normally compared to WT mice and do not show gross symptoms of autoimmunity (data not shown). Moreover, c-Maf deficiency in Treg cells does not affect the frequency of CD4 and CD8 T cells in lymphoid organs (**Figures 30A-C**), nor does it affect the activation profile of CD4 and CD8 T cells in the spleen and mLN (**Figures 30D-G**). This indicates that c-Maf deficiency in Treg cells

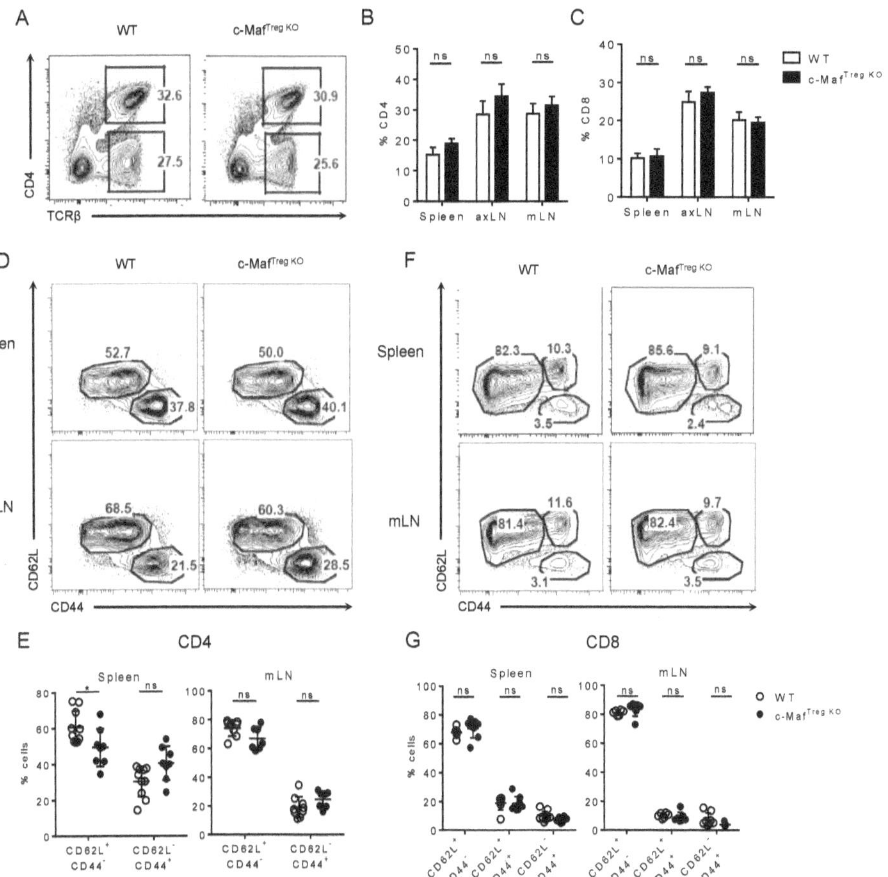

Figure 30. **c-Maf deficiency in Treg cells does not lead to the development of systemic immunity.** (A, D, F) Representative flow cytometry expression profiles of CD4 versus TCRβ in splenic live cells (A) and CD62L versus CD44 in CD4 T cells (D) or CD8 T cells (F) in the spleen and mLN of WT and c-Maf$^{Treg KO}$ mice. (B, C) Histograms show the frequency of CD4 (B) or CD8 (C) T cells in the spleen, axLN, and mLN of WT and c-Maf$^{Treg KO}$ mice. (E, G) Individual dots show the frequency of CD62L$^+$ CD44$^-$, CD62L$^-$ CD44$^+$, and CD62L$^+$ CD44$^+$ cells in CD4 (E) and CD8 (G) T cells in the spleen and mLN of WT and c-Maf$^{Treg KO}$ mice. Results are representative of at least three independent experiments; histograms represent the mean ± SD of at least three individual mice. Difference between groups is determined by a two-way ANOVA. ns: non-significant; *p < 0.05. (mLN: mesenteric lymph nodes)

does not disrupt global T cell homeostasis on a systemic level.

To determine whether c-Maf is required for the differentiation of Treg cells, we compared the frequency of Treg cells between WT and c-Maf$^{Treg\ KO}$ mice. Treg cells were able to differentiate in the absence of c-Maf, indicating that c-Maf is not required for the differentiation of Treg cells (**Figure 31A**), though c-Maf$^{Treg\ KO}$ mice exhibited higher frequencies of Treg cells, particularly in the intestine. c-Maf-deficient Treg cells showed a decreased expression of Foxp3 in lymphoid organs but not in the mLN or siLP (**Figure 31B**).

As previously determined, c-Maf expression is enriched among effector Treg cells, suggesting that c-Maf might be required for the activation of Treg cells, or simply act as an activation marker. To assess whether c-Maf expression is required for the activation of Treg cells, we compared the expression of CD62L and CD44 in Treg cells in WT and c-Maf-deficient Treg cells by flow cytometry. We observed a slight increase in the frequency of CD62L$^-$ CD44$^+$ effector Treg cells in the spleen of c-Maf$^{Treg\ KO}$ mice compared to WT mice (**Figures 31C-E**), associated with an equivalent decrease in CD62L$^+$ CD44$^-$ central Treg cells. The activation profile of Treg cells in mLN was not affected. Wheaton et al. similarly reported that Treg cell activation was not altered in absence of c-Maf[305], opposing a major role for c-Maf in the activation of Treg cells. To assess the suppressive capacity of c-Maf-deficient Treg cells, we conducted a classical *in vitro* suppression assay. Naïve CD4 T cells were activated *in vitro* by anti-CD3 stimulation, in presence of APCs, leading to cell proliferation. Co-culturing proliferating naïve T cells with Treg cells allowed us to measure Treg suppression of T cell proliferation. c-Maf-deficient Treg cells, purified from the spleen or mLN, were able to suppress conventional T cell proliferation similarly to WT Treg cells (**Figures 31F, 31G**). Taken together, our results show that c-Maf is not required for the differentiation, activation, and *in vitro* suppression of Treg cells, and that c-Maf deficiency does not alter the maintenance of central tolerance by Treg cells.

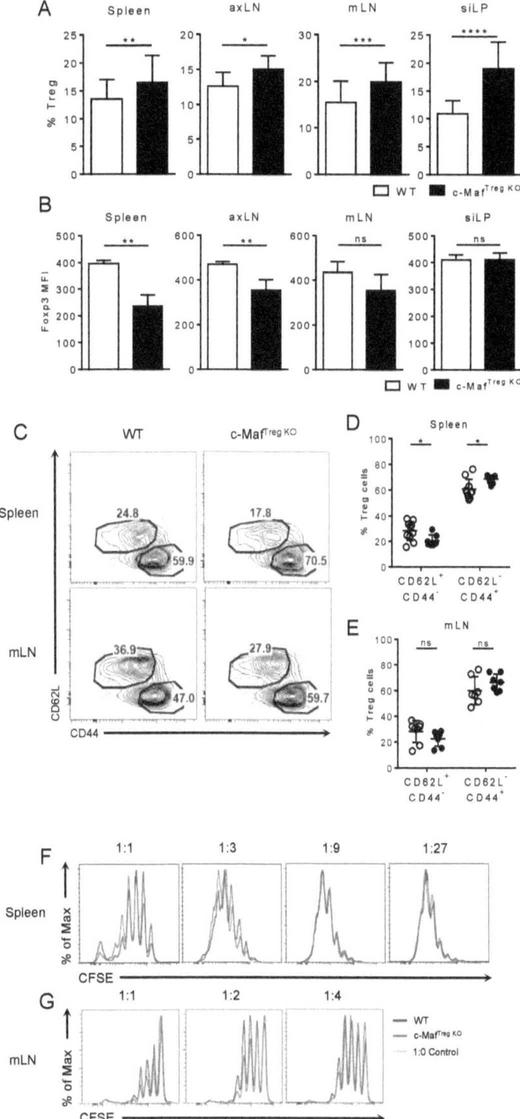

Figure 31. **c-Maf deficiency does not hinder Treg differentiation, activation, or *in vitro* suppression.** (A, B) Frequency of Treg cells (gate CD4+) and MFI of Foxp3 (gate Foxp3+) in the indicated organs of WT and c-Maf^Treg KO mice. (C – E) Representative flow cytometry expression profiles of CD62L versus CD44 in Treg cells (C), frequency of cTreg (CD62L+ CD44-) and eTreg (CD62L- CD44+) cells (D, E) in the spleen and mLN of WT and c-Maf^Treg KO mice. (F, G) Histograms showing CFSE staining profiles of conventional CD4 T cells in a Treg suppression assay *in vitro*, with Treg cells sorted from the spleen (F) or mLN (G) of WT and c-Maf^Treg KO mice. Histograms represent the mean ± SD of five individual mice. Difference between groups is determined by an unpaired t test (A), a Mann–Whitney test for two-tailed data (B), or a two-way ANOVA (D, E). ns: non-significant; *p < 0.05; **p < 0.01; ***p < 0.001; ****p < 0.0001. (axLN: axillary lymph nodes; mLN: mesenteric lymph nodes; siLP: small intestine lamina propria).

3.1.3. c-Maf-expressing Treg cells control homeostatic Th17 intestinal responses

While c-Maf$^{Treg\ KO}$ mice did not show signs of systemic auto-immunity, 50% of them spontaneously developed a rectal prolapse before 10 weeks of age (**Figures 32A, 32B**). Rectal prolapses are an external sign of colitis. They occur when intestinal inflammation provokes a loosening of the connective tissue attachments of the rectal mucosa, thus allowing the tissue to prolapse through the anus. In mice, they are often observed in immunodeficient mice, such as *Il10$^{-/-}$* mice which develop intestinal inflammation[422], or upon infection with pathobionts, such as *Helicobacter spp*[423]. Given the potential protective effect of a healthy microbiota on colitis, it is important to say that this observation was made in a facility where WT and c-Maf$^{Treg\ KO}$ mice were co-housed and where microbiota was exchanged between groups, due to the coprophagic behavior of mice. It is also important to note that the animal facility harboring c-Maf$^{Treg\ KO}$ mice does not contain *Helicobacter hepaticus*. To assess the level of inflammation observed in c-Maf$^{Treg\ KO}$ mice, we histologically scored colon tissue samples from WT and c-Maf$^{Treg\ KO}$ mice. While these results were not statistically significant, histological analysis of the colon lamina propria showed that c-Maf$^{Treg\ KO}$ mice tended to exhibit higher colitis scores at steady state (**Figures 32C, 32D**) compared to WT mice. c-Maf deficiency in Treg cells thus leads to the apparition of a spontaneous, low grade, intestinal inflammation, indicating that c-Maf is required for Treg cells to control homeostatic intestinal responses.

To characterize the intestinal inflammation observed in c-Maf$^{Treg\ KO}$ mice, we looked at intestinal cytokine expression by analyzing the mRNA expression of cytokines in total mLN of WT and c-Maf$^{Treg\ KO}$ mice. mLN are involved in the induction of intestinal immune responses. A disruption in the balance of pro- and anti-inflammatory cytokines in mLN is therefore likely to lead to inflammation at intestinal effector sites. We observed an increase in the expression of cytokines IL-17A, IL-22, and TNF-α (**Figure 33A**). The cytokine levels of other type 1 (IFN-γ, IL-12), type 2 (IL-4, IL-33), and type 3 (IL-1β, IL-6) cytokines in the mLN of c-Maf$^{Treg\ KO}$ mice were not affected. In addition to the increased Th17-associated cytokines, we observed a decrease in IL-10 mRNA, a cytokine associated with the suppression of immune responses (**Figure 33A**). This increase in pro-inflammatory

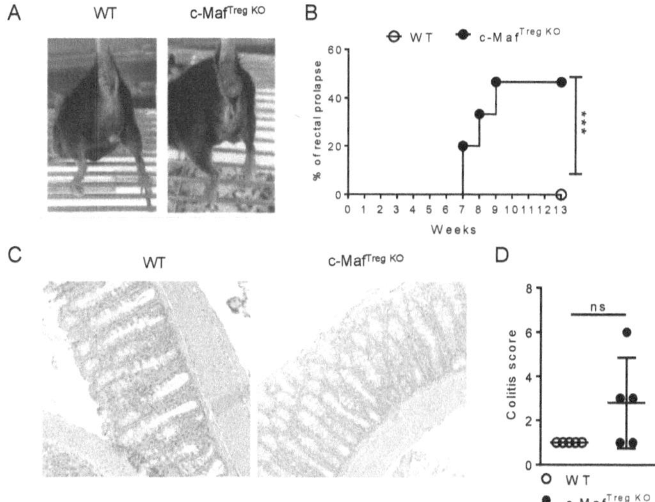

Figure 32. **c-Maf^Treg KO mice spontaneously develop intestinal inflammation.** (A) Rectal prolapse appears in c-Maf^Treg KO mice. (B) Incidence of rectal prolapse in WT and c-Maf^Treg KO mice. (C) HE staining of colon tissue from WT and c-Maf^Treg KO mice at steady state. (D) Colitis score of WT and c-Maf^Treg KO mice. Histograms represent the mean ± SD of at least five individual mice. Difference between groups is determined by a Mantel-Cox test (B) or a Wilcoxon signed rank test (D). ns: non-significant; ***p < 0.001

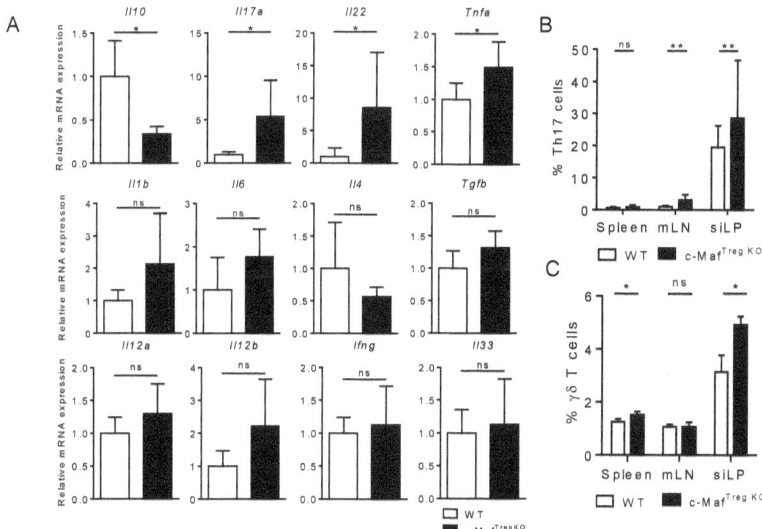

Figure 33. **c-Maf^Treg KO mice show an exacerbated Th17 intestinal response.** (A) Histograms show mRNA expression of cytokine genes in total WT or c-Maf^Treg KO mLN relative to RPL32. (B, C) Frequency of Th17 (RORγt⁺) cells among Tconv cells (B); γδ T cells among CD3⁺ cells (C) in the indicated organs of WT and c-Maf^Treg KO mice. Results are representative of at least three independent experiments; histograms represent the mean ± SD of at least five individual mice. Difference between groups is determined by a Mann–Whitney test for two-tailed data (A, D) and an unpaired t test (B). ns: non-significant; *p < 0.05; **p < 0.01. (mLN: mesenteric lymph nodes; siLP: small intestine lamina propria)

cytokines was associated with an increased frequency of Th17 cells in the mLN and siLP of c-Maf[Treg KO] mice compared to WT mice (**Figure 33B**). This observation has been also reported by others[307–309]. The frequency of Th1 and Th2 cells was not altered between WT and c-Maf[Treg KO] mice (data not shown). In addition, Neumann et al. reported that Th17 cells express higher levels of IL-17A[+] in c-Maf[Treg KO] mice[309]. These results suggest that Treg cells require c-Maf to suppress Th17 cell intestinal responses.

Using a broad panel of cell surface markers, we compared the immune cell composition of the spleen, mLN, and siLP of WT and c-Maf[Treg KO] mice, focusing on innate cell types. We observed that the frequency of γδ T cells found in the siLP of c-Maf[Treg KO] mice was increased (**Figure 33C**). The frequency of other immune cell subsets was not altered between WT and c-Maf[Treg KO] mice (data not shown). These observations indicate that c-Maf is required for Treg cells to control intestinal Th17 inflammation at steady state.

3.1.4. c-Maf is not required for Treg cells to control acute intestinal inflammation

We wanted to assess the role of c-Maf in the control of other types of intestinal inflammation by Treg cells. We tested the susceptibility of c-Maf[Treg KO] mice to different models of acute intestinal inflammation. We used three different models of acute inflammation: dextran sulfate sodium (DSS) ingestion, anti-CD3 mAb intraperitoneal injection, and *Toxoplasma gondii* oral infection. Despite important differences in the mechanisms underlying intestinal inflammation in these three models, they all rely on Treg cells for the resolution of inflammation. WT and c-Maf[Treg KO] mice showed similar survival and/or weight loss in the three models used (**Figures 34A-E**). c-Maf-deficiency in Treg cells did not influence the frequency of Th1 cells or activated conventional T cells after toxoplasmosis (**Figure 34F**). A recent study showed that T-bet expression in Treg cells is necessary to control Th1 inflammation during toxoplasmosis, but not to control the infection[424]. While this subset is not observed at steady state, we saw an increase in the frequency of T-bet[+] Treg cells after *Toxoplasma* infection. The frequency of T-bet[+] Treg cells was not altered in c-Maf[Treg KO] mice, suggesting that c-Maf is not necessary for the differentiation of this subset. These results show that c-Maf expression in Treg cells is not

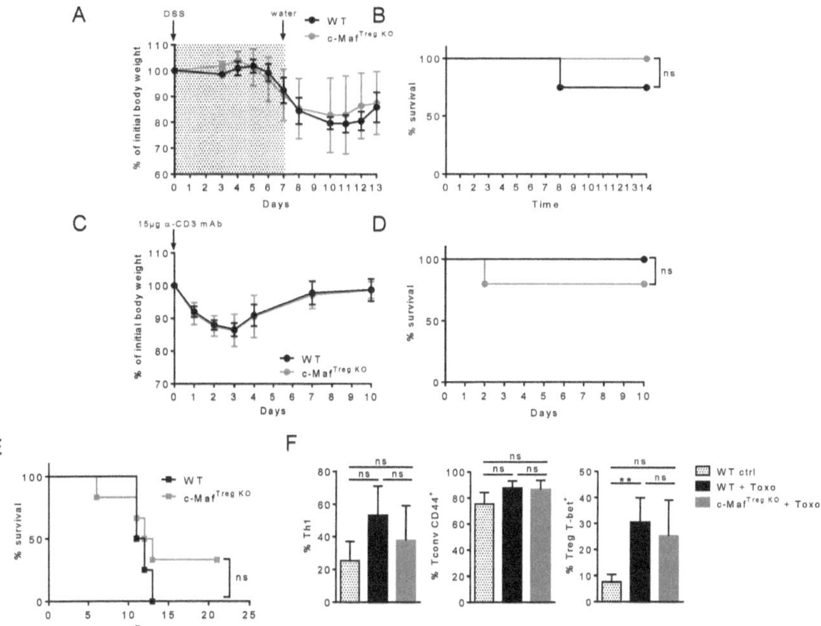

Figure 34. **c-Maf is not required for Treg cells to control acute intestinal inflammation.** (A, B) WT and c-Maf[Treg KO] mice were treated with DSS (2% in drinking water) for 7 days, (C, D) injected with 15μg anti-CD3 mAb, or (E, F) infected with *Toxoplasma gondii* (10 cysts, oral gavage). (A, C) Weight loss in percentage of initial body weight. (B, D, E) Survival curves. (F) Frequency of Th1, CD44+ Tconv, and T-bet+ Treg cells in siLP. Results are representative of at least two independent experiments. Graphs represent the mean ± SD of at least four individual mice. Difference between groups is determined by a two-way ANOVA (A, C, F) or a Mantel-Cox test (B, D, E). ns: non-significant; **p < 0.01

required for the control of acute intestinal inflammation, nor for the control of Th1 inflammation in particular.

3.1.5. c-Maf controls the acquisition of RORγt and IL-10 expression by intestinal Treg cells

Intestinal Treg cells form a complex population of cells that develop distinct functional phenotypes which preferentially target certain immune threats. Three notable intestinal Treg populations have been identified, on the basis of GATA3 and RORγt expression. We investigated the impact of c-Maf deficiency on the intestinal Treg phenotype by analyzing the expression of transcription factors GATA3 and RORγt among Treg cells. In WT mice, RORγt$^+$ Treg cells make up around 10% of Treg cells in mLN and 30% in siLP, while GATA3$^+$ Treg cells form 20% of Treg cells in both organs (**Figure 35A**). c-Maf$^{Treg KO}$ mice exhibit a 10-fold decrease in the frequency of RORγt$^+$ Treg cells, both in mLN and siLP (**Figure 35B**), while the frequency of GATA3$^+$ Treg cells is unaffected (**Figure 35C**). This indicates that c-Maf is required for the differentiation of RORγt$^+$ but not GATA3$^+$ Treg cells. RORγt$^+$ Treg cells form the highest IL-10 producing Treg subset in the intestine (**Figure 35D**). Coherently with the loss of the RORγt$^+$ Treg subset, c-Maf deficiency in Treg cells results in a decreased IL-10 expression by intestinal Treg cells (**Figure 35E**). While Wheaton et al. argue that the loss of IL-10 expression in Treg cells results directly from the loss of RORγt$^+$ Treg cells[305], our results show that a decrease in IL-10 expression can be observed in both the RORγt$^+$ (**Figure 35F**) and the RORγt$^-$ subset (**Figure 35G**), suggesting a role for c-Maf in the control of IL-10 expression by intestinal Treg cells. Similarly, Neumann et al. reported that, in absence of c-Maf in Treg cells, IL-10 expression was decreased among the Helios$^+$ and Helios$^-$ intestinal Treg cell subsets[309]. Taken together, our results show that c-Maf endows intestinal Treg cells with the ability to specialize into RORγt-expressing cells and to produce the anti-inflammatory cytokine IL-10. It is important to note that RORγt$^+$ Treg cells and IL-10 expression by Treg cells have both been identified as important for the control of Th17 immune responses[217,269]. In the absence of c-Maf, Treg cells are unable to upregulate RORγt and IL-10 expression, and are therefore unable to control steady state Th17 cell responses, leading to increased type 3 cytokine production and the development of colitis.

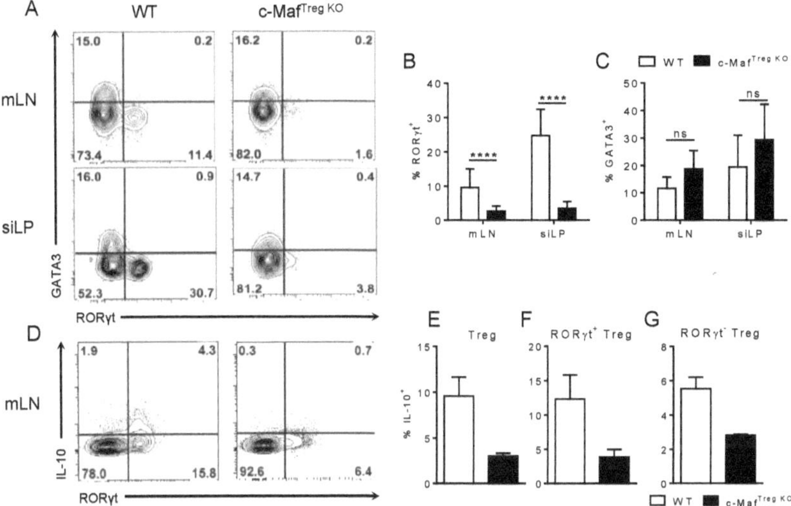

Figure 35. **c-Maf is required for the differentiation of RORγt+ Treg cells and IL-10 expression by Treg cells.** (A - C) Expression profiles of GATA3 versus RORγt (A) and frequency of RORγt+ (B) or GATA3+ (C) cells among Treg cells (gate CD4+ Foxp3+) from mLN and siLP of WT and c-Maf$^{Treg\ KO}$ mice. (D – G) Expression profiles of IL-10 versus RORγt among Treg cells (D) and frequency of IL-10+ cells among Treg (E), RORγt+ Treg (F), and RORγt- Treg (G) cells in mLN of WT and c-Maf$^{Treg\ KO}$ mice (preliminary experiment). Histograms represent the mean ± SD of at least six individual mice (B, C). Difference between groups is determined by an unpaired t test (B, C). ns: non-significant; ****p < 0.0001. (mLN: mesenteric lymph nodes, siLP: small intestine lamina propria).

3.1.6. c-Maf-expressing Treg cells promote the development of colorectal cancer during inflammation

In the context of cancer, Treg cells have a deleterious effect as they inhibit anti-tumor immunity, thus promoting tumor development and progression. Indeed, a high infiltration of Treg cells in the tumor microenvironment (TME) is associated with poor survival in various types of cancer, including melanoma and non-small-cell lung, gastric, and ovarian cancers[425,426]. Similar to the intestine, the TME is enriched in effector Treg cells, which express high levels of immunosuppressive molecules[427]. The high levels of c-Maf expression in effector Treg cells prompted us to look at the expression of c-Maf in tumor-infiltrating Treg cells by flow cytometry. Most tumor-infiltrating lymphocytes found in ectopic models of thymoma (EG7-OVA) and colon carcinoma (MC38) expressed c-Maf (**Figure 36A**). In EG7-OVA tumors, while Treg cells expressed low levels of c-Maf in non-draining lymph nodes, Treg cells upregulated c-Maf expression in the draining lymph nodes (**Figure 36B**). In the tumor, the majority of Treg cells have acquired c-Maf expression. This raises two hypotheses as to the site of c-Maf induction in tumor-infiltrating Treg cells. Treg cells that have upregulated c-Maf in lymph nodes could become primed to migrate to the tumor, which would explain the accumulation of c-Maf$^+$ Treg cells in the tumor. Alternatively, the tumoral environment could be directly responsible for the induction of c-Maf expression. Indeed, tumor cells have been shown to express high levels of TGF-β, which is involved in the induction of c-Maf expression in Th17 cells[428]. Due to the high expression of c-Maf in tumor-infiltrating lymphocytes, we wanted to determine the role of c-Maf in the control of tumor growth by Treg cells. To do so, we compared the growth of ectopic EG7-OVA and MC38 tumors in WT and c-Maf$^{\text{Treg KO}}$ mice. Despite the high expression levels of c-Maf among tumor-infiltrating Treg cells, the deletion of c-Maf in Treg cells did not alter the growth of ectopic EG7-OVA and MC38 tumors (**Figures 36C, 36D**). This suggests that, while c-Maf seems to be a marker of activated tumor-infiltrating Treg cells, the expression of c-Maf is not required for the control of ectopic tumor growth by Treg cells.

In colon cancer, the effect of Treg cells over cancer development can be twofold. On the one hand, Treg cells can suppress anti-tumor immunity and therefore promote tumor

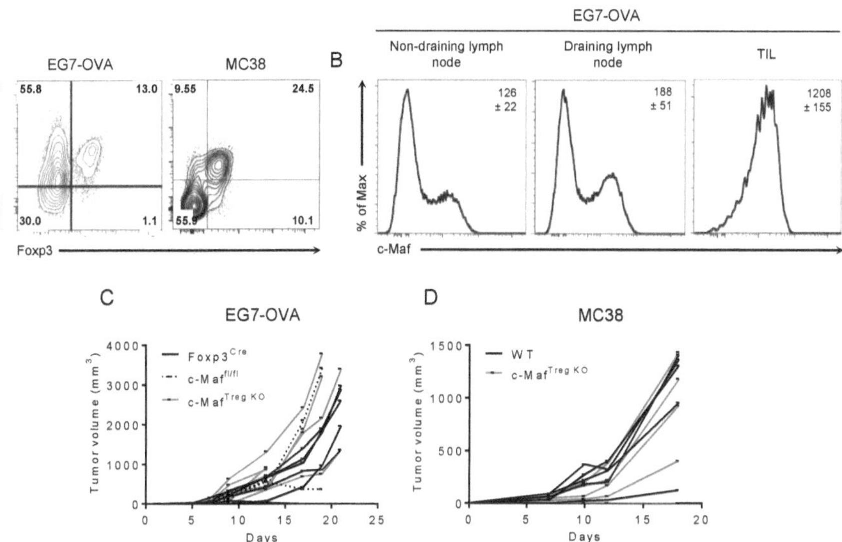

Figure 36. **c-Maf is widely expressed by tumor-infiltrating Treg cells but not required for the control of ectopic tumor growth.** (A) Representative flow cytometry expression profiles of c-Maf versus Foxp3 (gate CD4+) in the TILs isolated from EG7-OVA or MC38 tumors. (B) Histograms showing the expression of c-Maf in Treg cells in the indicated tissues. (C, D) Tumor growth of ectopic EG7-OVA (C) and MC38 (D) tumor cells in WT and c-Maf$^{Treg\ KO}$ mice. Results are representative of at least two independent experiments.

development. On the other hand, intestinal inflammation, and particularly chronic inflammation, promotes colon cancer development. Indeed, IBD patients show an increased incidence of colon cancer[429]. In this instance, the suppressive role of Treg cells decreases inflammation, and thus reduces the incidence of colon cancer. Our previous results identified a role of c-Maf in the control of intestinal inflammation by Treg cells at steady state. To investigate the role of c-Maf in the control of cancer development by Treg cells in the intestine, we used a model of colitis-induced colon cancer and compared polyp apparition in WT and c-Maf$^{Treg\ KO}$ mice. Mice were injected intraperitoneally with the carcinogen azoxymethane (AOM) on day 0, then, on day 3, treated with three 5-day courses of DSS, separated by two weeks of normal water (**Figure 37A**). AOM/DSS-treated mice develop polyps in the distal part of the colon (**Figure 37B**). c-Maf$^{Treg\ KO}$ mice develop lower numbers of polyps after AOM/DSS treatment, compared to WT mice (**Figures 37B, 37C**). In addition, polyps in c-Maf$^{Treg\ KO}$ mice show a decreased tumor score compared to polyps found in WT mice (**Figure 37D**). c-Maf$^{Treg\ KO}$ mice also exhibit an increased colitis score after AOM/DSS treatment compared to WT mice (**Figures 37E, 37F**). This increased inflammation is in line with the results observed at steady state. c-Maf deficiency in Treg cells leads to the development of fewer, less aggressive colon polyps, and to a higher level of intestinal inflammation.

We then characterized the immune response observed in c-Maf$^{Treg\ KO}$ after AOM/DSS treatment. Our preliminary results show that, similar to what was observed at steady state, c-Maf$^{Treg\ KO}$ mice show an increase in the frequency of Th17 cells (**Figures 38A, 38B**) and a decrease in the frequency of RORγt$^+$ Treg cells (**Figures 38A, 38C**) in the siLP after AOM/DSS treatment, leading to an increased Th17/Tr17 ratio in the siLP (**Figure 38D**). Analysis of the mRNA expression of cytokines showed that, while cytokine levels were not altered in the siLP, IL-10 expression was decreased and IL-17A expression was increased in the cLP of c-Maf$^{Treg\ KO}$ mice after AOM/DSS treatment (**Figure 38E**). This led to an increase in the IL-17A/IL-10 ratio in the cLP of c-Maf$^{Treg\ KO}$ mice (**Figure 38F**), indicative of an uncontrolled Th17 inflammation. Other cytokines such as IFN-γ, IL-6, IL-1β, TNF-α, and TGF-β were not altered. It is possible that IL-22 is also increased in the cLP of c-Maf$^{Treg\ KO}$ mice, but these preliminary results did not show statistical significance and need to be reproduced. Taken together, these results show that, while they favor the

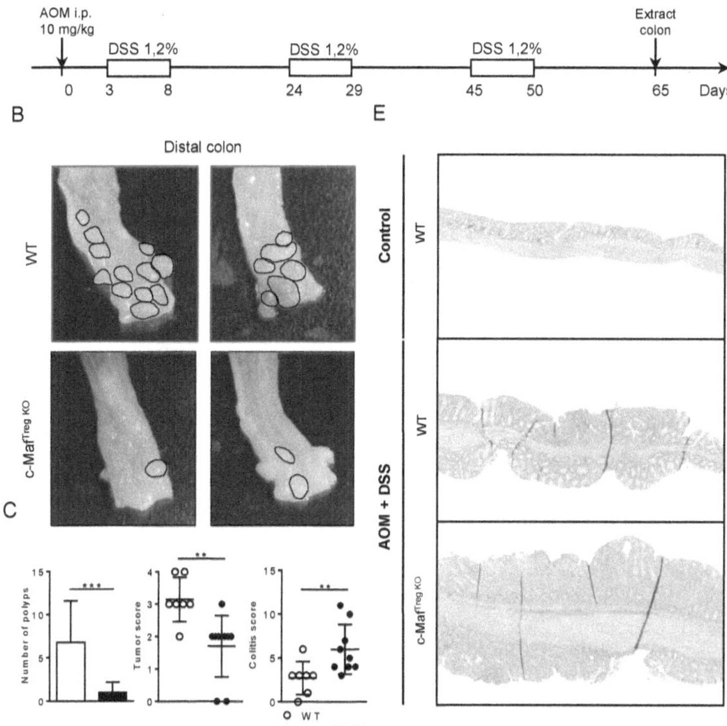

Figure 37. **c-Maf-deficiency in Treg cells protects mice from colitis-associated colon cancer.** (A) To induce colitis-associated colon cancer, mice were injected intraperitoneally with azoxymethane (10mg/kg) on day 0, followed by three courses of DSS (5 days, 1.2% in drinking water), and sacrificed on day 65. (B) Distal colon of WT and c-Maf^Treg KO mice. Histograms show the number of polyps (C) and tumor score (D) of WT and c-Maf^Treg KO mice. HE staining (E) and colitis score (F) of transversal colon sections from control and AOM/DSS treated WT and c-Maf^Treg KO mice. Results are representative of at least three independent experiments. Histograms represent the mean ± SD of at least seven individual mice. Difference between groups is determined by a Mann–Whitney test for two-tailed data. **p < 0.01; ***p < 0.001.

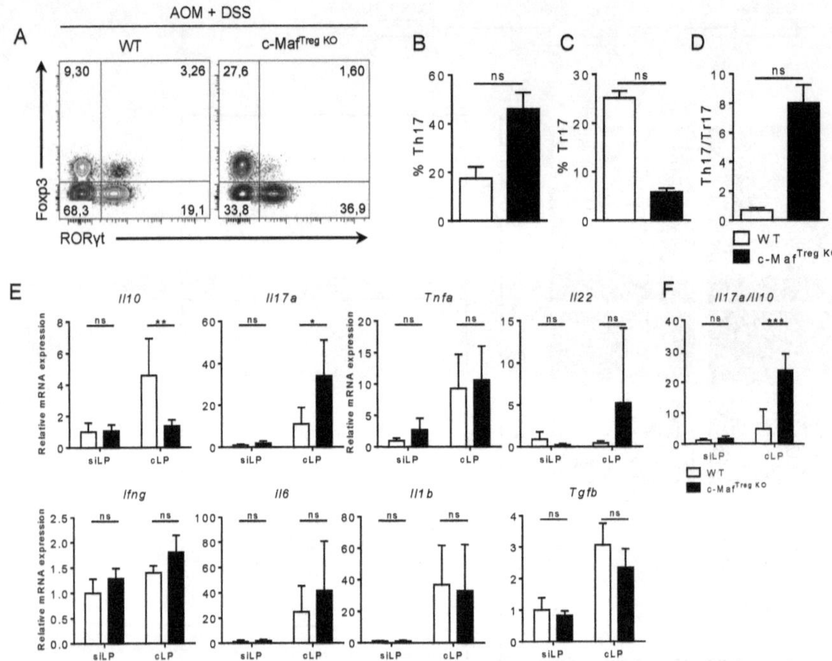

Figure 38. **c-Maf[Treg KO] mice have an exacerbated Th17 intestinal inflammation after AOM/DSS treatment.** Mice were treated as detailed in Fig. 37 A. (A – D) Expression profiles of Foxp3 versus RORγt (A; gate CD4+), frequency of Th17 (B), Tr17 (C; CD4+ Foxp3+ RORγt+) cells, and Th17/Tr17 ratio (D) in siLP of WT and c-Maf[Treg KO] mice. (E) mRNA expression of cytokine genes in total WT or c-Maf[Treg KO] siLP and cLP relative to RPL32. (F) *Il17a/Il10* mRNA ratio. Histograms represent the mean ± SD of at least four individual mice (preliminary experiment). Difference between groups is determined by a two-way ANOVA. ns: non-significant; *p < 0.05; **p < 0.01; ***p < 0.001.

maintenance of intestinal immune homeostasis, c-Maf$^+$ Treg cells promote the development of colon cancer during inflammation. These results will be discussed further in chapter 4.

Part 2: Molecular control over RORγt expression in Treg cells

3.2.1. Multiple environmental signaling pathways control the differentiation of RORγt-expressing regulatory T cells

Intestinal Treg cells differentiate into three specialized Treg subsets which control immune responses directed towards distinct threats. The mechanisms underlying the acquisition of each intestinal Treg phenotype and the regulation of these subsets, at steady state or in an inflammatory context, are still unknown. Among intestinal Treg subsets, RORγt[+] Treg cells are the object of a growing body of literature[268,269,301,303,430]. This subset is essential to the maintenance of gut homeostasis and to the tolerance to commensal microbiota. Recently, transcription factor c-Maf has emerged as a critical factor in the regulation of RORγt expression in Treg cells[305,307–309]. However, aside from c-Maf, the signaling pathways involved in the differentiation of RORγt[+] Treg cells, as well as their possible interplay with c-Maf in this process, are still incompletely defined.

The second part of this work focused on the molecular mechanisms involved in the expression of RORγt in intestinal Treg cells. This work was published in Frontiers of Immunology on January 8th 2020 as a first-author publication titled "Multiple environmental signaling pathways control the differentiation of RORγt-expressing regulatory T cells"[310]. The conclusions of this work are briefly summarized below and are followed by the full text of the article.

We first confirm that c-Maf is highly expressed in intestinal Treg cells and is required for RORγt expression (**Figures 1, 2**). These results have been extensively described in Part 1 of this manuscript. Using genetically invalidated mouse models, we then show that the signals derived from a complex microbiota, the IL-6/STAT3, and the TGF-β signaling pathways act as positive regulators of RORγt expression in Treg cells (**Figures 3, 4**). While these signals are required for RORγt expression, they are however dispensable for the induction of c-Maf in Treg cells (**Figures 3, 4**). Using an *in vitro* approach, we also show that ectopic expression of c-Maf does not rescue RORγt expression in STAT3-deficient Treg cells (**Figures 5, 6**). Finally, using a combination of *in vivo* and *in vitro* approaches, we show that Th1 inflammation, and particularly the IFN-γ/STAT1 pathway, acts as a negative regulator of RORγt expression in Treg cells (**Figure 7**).

Our data thus argue that a complex integrative network of signals finely tunes RORγt expression in Treg cells, where STAT1 might exert a dominant negative effect over STAT3.

ORIGINAL RESEARCH
published: 08 January 2020
doi: 10.3389/fimmu.2019.03007

Multiple Environmental Signaling Pathways Control the Differentiation of RORγt-Expressing Regulatory T Cells

[1] Laboratoire d'Immunobiologie, Université Libre de Bruxelles, Brussels, Belgium, [2] Diabetes Center, University of California, San Francisco, San Francisco, CA, United States, [3] Sean N. Parker Autoimmune Research Laboratory, University of California, San Francisco, San Francisco, CA, United States, [4] Institute for Medical Immunology (IMI), Université Libre de Bruxelles, Gosselies, Belgium

RORγt-expressing Tregs form a specialized subset of intestinal CD4[+] Foxp3[+] cells which is essential to maintain gut homeostasis and tolerance to commensal microbiota. Recently, c-Maf emerged as a critical factor in the regulation of RORγt expression in Tregs. However, aside from c-Maf signaling, the signaling pathways involved in the differentiation of RORγt[+] Tregs and their possible interplay with c-Maf in this process are largely unknown. We show that RORγt[+] Treg development is controled by positive as well as negative signals. Along with c-Maf signaling, signals derived from a complex microbiota, as well as IL-6/STAT3- and TGF-β-derived signals act in favor of RORγt[+] Treg development. Ectopic expression of c-Maf did not rescue RORγt expression in STAT3-deficient Tregs, indicating the presence of additional effectors downstream of STAT3. Moreover, we show that an inflammatory IFN-γ/STAT1 signaling pathway acts as a negative regulator of RORγt[+] Treg differentiation in a c-Maf independent fashion. These data thus argue for a complex integrative signaling network that finely tunes RORγt expression in Tregs. The finding that type 1 inflammation impedes RORγt[+] Treg development even in the presence of an active IL-6/STAT3 pathway further suggests a dominant negative effect of STAT1 over STAT3 in this process.

Keywords: Treg subsets, transcription factors, RORγt, cell differentiation, signal transduction, c-Maf

OPEN ACCESS

Specialty section:
This article was submitted to
T Cell Biology,
a section of the journal
Frontiers in Immunology

Received: 15 October 2019
Accepted: 09 December 2019
Published: 08 January 2020

Citation:
Hussein H, Denanglaire S, Van Gool F,
Azouz A, Ajouaou Y, El-Khatib H,
Oldenhove G, Leo O and Andris F
(2020) Multiple Environmental
Signaling Pathways Control the
Differentiation of RORγt-Expressing
Regulatory T Cells.
Front. Immunol. 10:3007.
doi: 10.3389/fimmu.2019.03007

INTRODUCTION

CD4 T cells expressing transcription factor forkhead box P3 (Foxp3) constitute a regulatory lineage of T cells (Treg) which maintains immune tolerance against self-antigens and prevents tissue destruction consequent to excessive immune responses. Tregs can be generated in the thymus from developing CD4[+] thymocytes (tTregs) or can result from the differentiation of mature T cells in the periphery (pTregs) (1–3). Recent findings indicate that, similarly to conventional helper T cells, Tregs are phenotypically and functionally heterogeneous. Distinct Treg populations adopt specialized phenotypes through the co-expression of Foxp3 and lineage-defining transcription factors such as PPARγ, BCL6, or RORγt in response to tissue- or inflammatory-driven signals (3).

In particular, RORγt[+] Tregs are found in the intestinal tissue of naïve mice. Signals deriving from a complex microbiota as well as STAT3 signaling are necessary to their presence in the intestinal compartment (4, 5). This subset of Tregs has been shown to protect efficiently from intestinal immunopathology in different colitis models (4–7) and to mediate immunological tolerance to the gut pathobiont *Helicobacter hepaticus* (8). A recent study showed that thymic-derived Tregs also expressed RORγt in lymph nodes following immunization and were able to protect mice from Th17 cell-mediated CNS inflammation (9). Although RORγt expression can be acquired by ex-Tregs during pathogenic Th17 conversion (10), RORγt-expressing Tregs mostly represent a Treg lineage which participates in the immunological tolerance of barrier tissues and protects from autoimmunity (4–9). RORγt$^+$ Tregs can also develop in the tumor microenvironment where they hinder anti-tumor immunity, thus revealing a double-edged function of this Treg subset in immune homeostasis (11). However, despite their importance in physiological and pathological immune responses, the factors driving RORγt$^+$ Treg differentiation are still incompletely defined.

Recent studies reported that transcription factor c-Maf promotes the differentiation of intestinal RORγt$^+$ Tregs (8, 12, 13). Coincidentally, transcriptomic studies conducted on Tregs originating from different tissues revealed a strong enrichment for c-Maf in the intestinal compartment (4). Transcription factor c-Maf, a member of the AP-1 family of basic region/leucine zipper transcription factors, is expressed by distinct CD4$^+$ T cell subsets, including Th17, Th2, Tfh, and Tr1 cells, and is thought to regulate the expression of IL-10, IL-4, and IL-21 through the transactivation of their promoters, downstream of Batf, ICOS, and STAT3 signaling (14–18). Thus, and similarly to what has been previously described for Th17 cells (19), expression of RORγt in Tregs is c-Maf-dependent. However, unlike RORγt expression, which is restricted to gut-associated Tregs in naïve mice, c-Maf is expressed by a wider proportion of Tregs found in distinct organs. Of note, high levels of c-Maf are found in a subset of splenic CD44$^+$ CD62L$^-$ effector Tregs driven by ICOS signaling (12). The partial overlapping of RORγt and c-Maf expression along with the presence of a substantial population of c-Maf$^+$ RORγt$^-$ Tregs in lymphoid organs therefore suggests that c-Maf is not sufficient *per se* to drive RORγt$^+$ Treg cell differentiation and supports the existence of complementary signaling pathways.

Herein, extensive analysis of the lymphoid organs and tissues of genetically invalidated mice or mice harboring an altered microbiota revealed that, well beyond the c-Maf/RORγt interplay, multiple signaling pathways cooperate to exert a tight control over RORγt expression in Tregs.

MATERIALS AND METHODS

Mice

C57BL/6 mice were purchased from Envigo (Horst, The Netherlands). c-Maf-flox mice (C. Birchmeier, Max Delbrück Center for Molecular Medicine, Berlin, Germany) were crossed with CD4-CRE mice (G. Van Loo, Ghent University, Ghent, Belgium) or FOXP3-CRE-YFP mice which were developed by Rudensky (20) and kindly provided by A. Liston (KU Leuven, Leuven, Belgium). IL-6$^{-/-}$ mice were obtained from The Jackson Laboratory (Bar Harbor, ME, USA). STAT3-flox mice were kindly provided by S. Akira (Osaka University, Osaka, Japan); STAT1$^{-/-}$ mice by D.E. Levy (New York University School of Medicine, NYC, USA). Germ-free mice were obtained from the Ghent Germfree and Gnotobiotic mouse facility (Ghent University, Ghent, Belgium) and were compared to SPF control mice. c-Maf-flox, CD4-CRE, FOXP3-CRE-YFP, IL-6$^{-/-}$, STAT3-flox, STAT1$^{-/-}$ and germ-free mice were bred on a C57BL/6 background. Tgfbr2-flox mice (21) on a NOD background crossed with Foxp3-Cre mice (JAX 008694) were kindly provided by Q. Tang (University of California San Francisco, SF, USA) and were housed and bred at the UCSF Animal Barrier Facility.

All mice were used between 6 and 12 weeks of age. The experiments were carried out in compliance with the relevant laws and institutional guidelines and were approved by the Université Libre de Bruxelles Institutional Animal Care and Use Committee (protocol number CEBEA-4).

Antibodies, Intracellular Staining, and Flow Cytometry

The following monoclonal antibodies were purchased from eBioscience: CD278 (ICOS)-biotin, CD304 (Nrp1)-PerCP eF710, c-Maf-EF660, Foxp3-FITC, RORγt-PE, T-bet-PE; or from BD Biosciences: CD25-BB515, CD44-PECy7, CD4-A700, CD4-PB, CD62L-A700, GATA3-PE, RORγt-PECF594, STAT1 (pY701)-A488, STAT3 (pY705)-A647, streptavidin-PECy7.

Live/dead fixable near-IR stain (ThermoFisher) was used to exclude dead cells. For transcription factor staining, cells were stained for surface markers, followed by fixation and permeabilization before nuclear factor staining according to the manufacturer's protocol (FOXP3 staining buffer set from eBioscience). For phosphorylation staining, cells were fixed with formaldehyde and permeabilized with methanol before staining. Flow cytometric analysis was performed on a Canto II (BD Biosciences) or CytoFLEX (Beckman Coulter) and analyzed using FlowJo software (Tree Star).

Leukocyte Purification

After removal of Peyer's patches and mesenteric fat, intestinal tissues were washed in HBSS 3% FCS and PBS, cut in small sections and incubated in HBSS 3% FCS containing 2.5 mM EDTA and 72.5 μg/mL DTT for 30 min at 37°C with agitation to remove epithelial cells, and then minced and dissociated in RPMI containing liberase (20 μg/ml, Roche) and DNase (400 μg/ml, Roche) at 37°C for 30 min (small intestine) or 45 min (colon). Leukocytes were collected after a 30% Percoll gradient (GE Healthcare). Lymph nodes, thymus and spleens were mechanically disrupted in culture medium.

After anesthesia, mice were perfused with PBS. Liver and lung samples were digested with collagenase (200 U, Worthington) and DNase I (40 μg/mL, Roche) at 37°C and mechanically disrupted. Leukocytes were collected at the interphase of a 40/70% Percoll gradient.

Antibiotics Treatment

Wide spectrum antibiotics (ampicillin 1 g/L and neomycin 1 g/L, Sigma-Aldrich; vancomycin 0.5 g/L and metronidazole 1 g/L,

Duchefa) were added to the sweetened drinking water of mice treated with antibiotics for 3-4 weeks. Control mice were given sweetened drinking water in parallel.

Toxoplasma Infection

ME-49 type II *T. gondii* was kindly provided by Dr. De Craeye (Institut Scientifique de Santé Publique, Belgium) and was used for the production of tissue cysts in C57BL/6 mice, which were inoculated 1-3 months previously with three cysts by gavage. Animals were killed, and the brains were removed. Tissue cysts were counted and mice were infected with 10 cysts by intragastric gavage. Mice were sacrificed at day 8 after infection.

T Cell Culture

Naïve CD4$^+$ T cells were purified from spleen of mice of indicated genotypes. CD4$^+$ T cells were positively selected from organ cell suspensions by magnetic-activated cell sorting using CD4 beads (MACS, Miltenyi) according to the product protocol, and then isolated as CD4$^+$ CD44^{lo}CD62L^{hi}CD25$^-$ or CD4$^+$ CD44^{lo} CD62L^{hi} YFP$^-$ by FACS. T cells were cultured at 37°C in RPMI supplemented with 5% heat-inactivated FBS (Sigma-Aldrich), 1% non-essential amino acids (Invitrogen), 1 mM sodium pyruvate (Invitrogen), 2 mM L-glutamin (Invitrogen), 500 U/mL penicillin/500 µg/ml streptomycin (Invitrogen), and 50 µM β-mercaptoethanol (Sigma-Aldrich).

To generate iTreg cells, cells were cultured 72 h in 24 or 96 well plates coated with 5 µg/mL anti-CD3 (BioXcell, 145-2c11) at 37°C for 72 h. The culture was supplemented with anti-CD28 (1 µg/mL, BioXcell, 37.51) and TGF-β alone (3 ng/ml, eBioscience), TGF-β and IL-2 (10 ng/mL, Peprotech), or TGF-β, IL-2 and IL-6 (10 ng/mL, eBioscience) for optimal iTreg cell polarization. Acetate (C2, 10 mM), propionate (C3, 0.5 mM), butyrate (C4, 0.125 mM), all from Sigma-Aldrich, and IFN-γ (10 and 100 ng/mL, Peprotech) were also used and added to the culture for the whole duration of the experiment.

Retroviral Infection

Platinum-E retroviral packaging cells (T. Kitamura, University of Tokyo, Tokyo, Japan) were transfected with a c-Maf encoding retroviral plasmid (pMIEG-c-Maf-IRES-GFP) or a control retroviral plasmid (pMIEG-IRES-GFP) to produce retrovirus-containing supernatants. 24 h after activation, naïve CD4 T cells polarized in presence of TGF-β, IL-2, and IL-6, as described above, were infected during a 90-min centrifugation with 1 mL retrovirus-containing supernatant and polybrene. Forty-eight hours later, infected cells were FACS-sorted based on GFP expression and were stimulated with anti-CD3 for 24 h (5 µg/mL, coated) before flow cytometry staining.

Treg Cell *in vitro* Suppression Assay

Naïve T cells with the phenotype CD4$^+$ CD44^{lo} CD62L^{hi} YFP$^-$ were isolated from the spleen of FOXP3-CRE-YFP mice by FACS after positive enrichment for CD4$^+$ cells using MACS LS columns (Miltenyi) and labeled with carboxyfluorescein diacetate succinimidyl ester (CFSE, ThermoFisher). Treg cells with the phenotype CD4$^+$ FOXP3-CRE-YFP$^+$ were isolated from the mesenteric lymph nodes of Foxp3-CRE-YFP or c-Maf^{TregKO} mice

by FACS. Splenocytes from wild-type B6 mice were depleted in T cells (anti-CD90.2 beads) using MACS LS columns (Miltenyi). 4×10^4 CFSE-labeled naïve T cells were cultured for 72 h with APCs (1×10^5) and soluble anti-CD3 (0.5 µg/mL) in the presence or absence of various numbers of Treg cells as indicated.

RT-qPCR

RNA was extracted using the TRIzol method (Invitrogen) and reverse transcribed with Superscript II reverse transcriptase (Invitrogen) according to the manufacturer's instructions. Quantitative real-time RT-PCR was performed using the SYBR Green Master mix kit (ThermoFisher). Primer sequences were as follow: *Rpl32* (F) ACATCGGTTATGGGAGCAAC; *Rrpl32* (R) TCCAGCTCCTTGACATTGT; *Il10* (F) CCTGGGTGAGAA GCTGAAGA; *Il10* (R) GCTCCACTGCCTTGCTCTTA; *Il17a* (F) ATCCCTCAAAGCTCAGCGTGTC; *Il17a* (R) GGGTCT TCATTGCGGTGGAGAG; *Il22* (F) CAGCAGCCATACATC GTCAA; *Il22* (R) GCCGGACATCTGTGTTGTTA; *Tnfa* (F) GCCTCCCTCTCATCAGTTCTA; *Tnfa* (R) GCTACGACGT GGGCTACAG; *Rorc* (F) TCTACACGGCCCTGGTTCT; *Rorc* (R) ATGTTCCACTCTCCTCTTCTCTTG.

ChIP-seq Data

Publicly available ChIP-seq data (GSE40918) for c-Maf and Stat3 was downloaded and mapped to the mm9 genome using Bowtie2 with sensitive-local predefined parameters. Resulted BAM files were converted to bigwig files and visualized by IGV genome browser.

Statistical Analysis

For unpaired data, statistical difference between groups was determined by an unpaired t test when the sample size was sufficient and both groups passed the normality test, and by a Mann-Whitney test for two-tailed data otherwise. Mutant and control groups did not always have similar standard deviations and therefore an unpaired two-sided Welch's t-test was used. For paired data, a paired t test was used. Error bars represent mean ± SD. No samples were excluded from the analysis.

RESULTS

c-Maf Is Highly Expressed in Intestinal Tregs and Is Required for RORγt Expression

We first investigated the expression of c-Maf and RORγt in Tregs found in distinct organs. In accordance with previous data (4, 5), we observed that RORγt$^+$ Tregs were mainly present in the small intestine and colon lamina propria and, to a lesser extent, in mesenteric lymph nodes (**Figures 1A,B**). Of note, a large proportion of Tregs (ranging from 20 to 85%) expressed c-Maf in all the examined organs, with a notable exception for the thymus (**Figures 1A,C**). In the intestine, both RORγt$^+$ and RORγt$^-$ Treg subsets expressed c-Maf, although expression levels were higher in the RORγt$^+$ compartment (**Figures 1A,D,E**). Strikingly, FACS analysis also revealed that a large fraction of c-Maf$^+$ Tregs do not express RORγt. This was observed in all the examined organs and was most evident in the small intestine lamina propria,

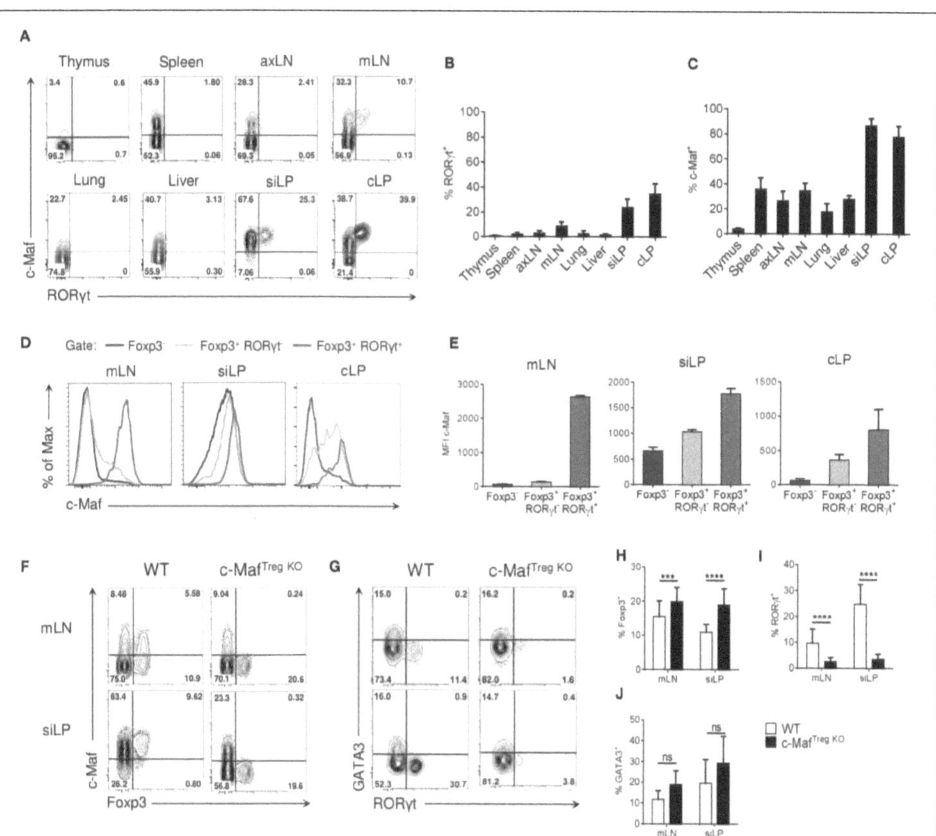

FIGURE 1 | Transcription factor c-Maf is required for the differentiation of RORγt$^+$ Tregs. **(A)** Representative flow cytometry expression profiles of c-Maf vs. RORγt of Treg cells in the indicated organs of naive C57BL/6 mice (gate CD4$^+$ Foxp3$^+$). **(B,C)** Histograms showing the frequency of RORγt$^+$ **(B)** or c-Maf$^+$ **(C)** Treg cells in the indicated organs. **(D,E)** Expression profile **(D)** and median of fluorescence intensity **(E)** of c-Maf among Foxp3$^-$, Foxp3$^+$ RORγt$^-$, and Foxp3$^+$ RORγt$^+$ CD4 T cells in the indicated organs. **(F,G)** Representative flow cytometry expression profiles of c-Maf vs. Foxp3 (F, gate CD4$^+$) and GATA3 vs. RORγt (G, gate CD4$^+$ Foxp3$^+$) from mLN and siLP of WT and c-Maf^{TregKO} mice. **(H–J)** Histograms showing the frequency of total Tregs **(H)**, RORγt$^+$ Tregs **(I)**, and GATA3$^+$ Tregs **(J)** in mLN and siLP of WT and c-Maf^{TregKO} mice. Results are representative of at least three independent experiments; histograms in **(B,C,E,H–J)** represent the mean ± SD of at least five individual mice. Difference between groups is determined by an unpaired t test **(H,I)** or a Mann-Whitney test for two-tailed data **(J)**. ***$p < 0.001$; ****$p < 0.0001$. axLN, axillary lymph nodes; mLN, mesenteric lymph nodes; siLP, small intestine lamina propria; cLP, colon lamina propria.

where two-thirds of the c-Maf$^+$ Tregs lacked RORγt expression (**Figure 1A**).

In lymphoid organs, c-Maf expression was mainly found among effector Tregs, characterized by the expression of high levels of ICOS and CD44 (**Figure S1**). In contrast to RORγt$^+$ Tregs, which were mostly of peripheral origin, c-Maf$^+$ Tregs were found both in Nrp1$^+$ and Nrp1$^-$ Tregs, suggesting that c-Maf$^+$ Tregs can be of thymic or peripheral origin (**Figures 2A,B**). Thymic c-Maf$^+$ Tregs were enriched in the spleen whereas

their peripheral counterparts were enriched in mesenteric lymph nodes and formed the majority of intestinal Tregs (**Figure 2C**).

Treg-specific ablation of c-Maf resulted in increased Treg proportions in the intestine and lymphoid organs (**Figures 1F,H** and **Figure S2A**). Despite the wide distribution of c-Maf-expressing Tregs in distinct organs, Treg-conditional ablation of c-Maf did not result in systemic autoimmune disease, nor did it disturb conventional and regulatory T cell homeostasis in lymphoid organs. c-Maf-deficient Tregs also retained their

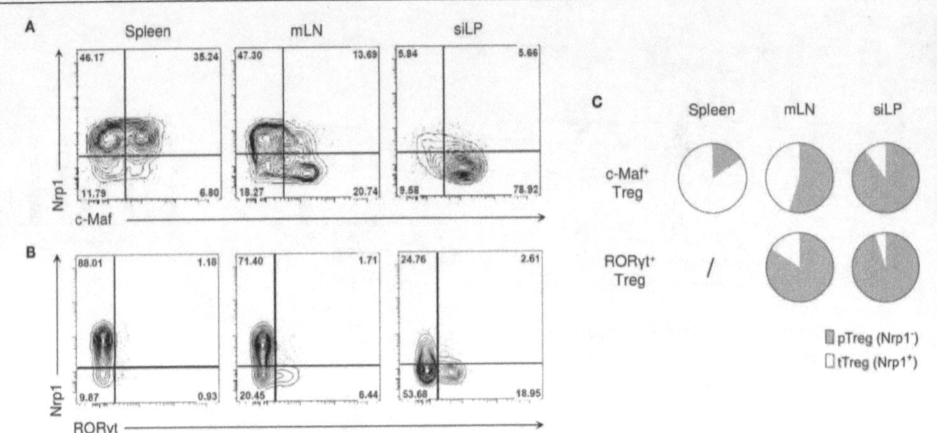

FIGURE 2 | c-Mat-expressing Tregs can be of thymic or peripheral origin. **(A,B)** Representative flow cytometry expression profiles of Nrp1 vs. c-Maf **(A)** or RORγt **(B)** of Treg cells in the indicated organs of naïve C57BL/6 mice (gate CD4⁺ Foxp3⁺ cells). **(C)** Pie charts show the relative frequencies of thymic (tTreg) and peripheral (pTreg) Treg cells among c-Maf⁺ and RORγt⁺ Treg subsets in the spleen, mLN and siLP. Results are representative of at least three independent experiments.

in vitro suppressive capacity (**Figure S2B** and data not shown). In agreement with previous reports (8, 12), c-Maf^{TregKO} mice spontaneously developed intestinal inflammation and showed a near complete loss of RORγt expression in Tregs (**Figures 1G,I** and **Figure S3**). They nevertheless expressed normal percentages of intestinal GATA3⁺ Tregs (**Figures 1G,J**).

Altogether, these data indicate that c-Maf is required for the differentiation of RORγt⁺ Tregs and that, contrary to RORγt, c-Maf expression is also found in a large proportion of effector thymic Tregs, located in distinct organs.

Complex Microbiota, STAT3 and TGF-β Signals Control RORγt but Are Dispensable for c-Maf Expression in Tregs

The presence of c-Maf⁺ RORγt⁻ Tregs in different organs and their distinct origins prompted us to further analyze the specific environmental signals responsible for the induction of c-Maf and RORγt expression by Tregs. IL-6/STAT3 and TGF-β signaling has been shown to drive RORγt⁺ and c-Maf expression in a variety of T cells (5, 22). Mice genetically invalidated for IL-6 (IL-6$^{-/-}$), STAT3 (Stat3$^{flox/flox}$-CD4^{CRE}) or TGF-β (Tgfbr2$^{flox/flox}$-Foxp3^{CRE}) signaling showed normal to even slightly increased frequencies of c-Maf⁺ Tregs, although RORγt expression was considerably decreased in all the aforementioned conditions (**Figure 3**; see also **Figure S4** for representative FACS plots). Thus, despite being necessary for the expression of RORγt by Tregs, c-Maf is not sufficient, and most likely cooperates with other signaling pathways to induce RORγt expression in Tregs.

RORγt⁺ Tregs are absolutely dependent on microbiota for their development and selected bacterial species or bacterial consortia can restore the expression RORγt in Tregs of germ-free mice (4, 5, 23). Analysis of germ-free and antibiotics-treated mice revealed that, in contrast to RORγt expression, which was lost in both cases, c-Maf expression by Tregs was only marginally affected in microbiota-deficient mice (**Figures 4A–D**).

Short chain fatty acids (SCFA) are gut microbiota-derived bacterial fermentation products, which include acetate, propionate, and butyrate, that regulate the size and function of the colonic Treg pool (24). Treg cells were induced *in vitro* with TGF-β in the absence or presence of acetate, propionate, or butyrate. In this experimental setting, SCFA did not affect or slightly decreased the differentiation of Foxp3⁺ Treg cells (**Figure S5**). Addition of SCFA to the culture medium induced a 4-5-fold upregulation of RORγt expression, while minimally affecting c-Maf expression (**Figures 4E,F**, left panels). In absence of c-Maf, intermediate levels of RORγt were induced in Tregs treated with SCFA (**Figure 4F**, right panel). Overall, these observations suggest that microbiota-derived products regulate RORγt expression in Tregs through both c-Maf-dependent and independent pathways.

STAT3 and c-Maf Control RORγt Expression in iTregs Through Partly Overlapping Pathways

In presence of TGF-β and IL-2, about 75% of *in vitro* activated naïve CD4 T cells differentiated into Tregs, as assessed by their Foxp3 expression. Addition of IL-6, a pro-Th17 cytokine, to the TGF-β/IL-2 cytokine cocktail decreased the frequency of induced Tregs but led to the differentiation of a population of double positive c-Maf⁺ RORγt⁺ induced Tregs (iTreg17 cells; **Figure 5A**, lower panels and **Figures 5B,C**). In agreement

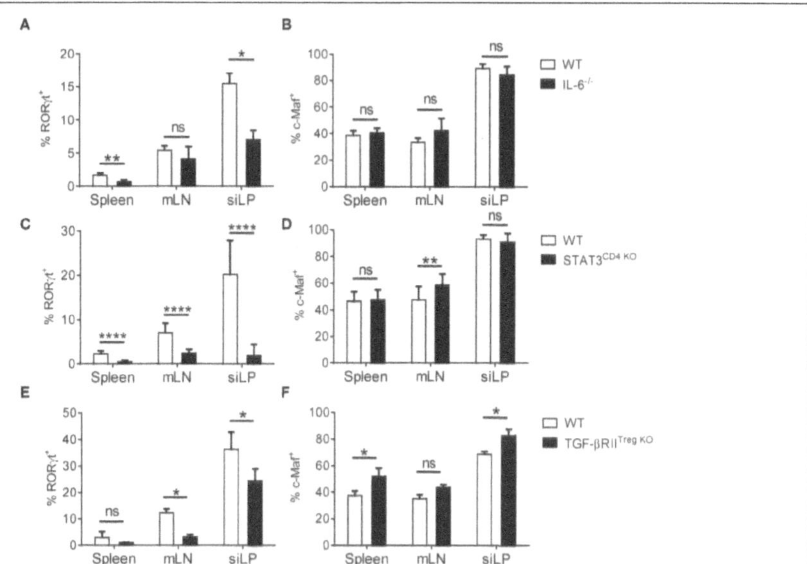

FIGURE 3 | IL-6/STAT3 and TGF-β signaling promote RORγt expression in Tregs independently of c-Maf. **(A–F)** Histograms show the frequency of RORγt⁺ **(A,C,E)** or c-Maf⁺ **(B,D,F)** cells among Treg cells (gate CD4⁺ Foxp3⁺) in indicated organs of WT and IL-6$^{-/-}$ **(A,B)**, STAT3^{CD4KO} **(C,D)** and TGF- βRII^{TregKO} **(E,F)** mice. Results are representative of at least three independent experiments; histograms represent the mean ± SD of at least four individual mice. Difference between groups is determined by an unpaired t test or a Mann-Whitney test for two-tailed data **(A,B,C** siLP, **D** mLN, **E,F)**. See **Figure S4** for representative dot plots. $^*p < 0.05$; $^{**}p < 0.01$; $^{****}p < 0.0001$.

with *in vivo* observations (**Figures 1G,I**), ablation of c-Maf expression led to a significant reduction of *in vitro* generated RORγt⁺ Tregs (**Figures 5D,E,H**). In the absence of STAT3, the expression of c-Maf and RORγt in iTreg17 was heavily decreased (**Figures 5F–I**). RT-qPCR analysis confirmed a transcriptional control of *Rorc* mRNA expression in these experimental settings (**Figure S6**). Despite showing residual c-Maf expression, STAT3 KO iTregs displayed a more severe down-regulation of RORγt than their c-Maf KO counterparts (90 vs. 50 %; **Figure 5J**), suggesting a prominent role of STAT3 in RORγt expression in this context. Analysis of STAT3 and c-Maf genome mapping from public ChIPseq databases (25) revealed that both transcription factors bind to the *Rorc* locus, albeit at distinct preferential sites (**Figure 5K**).

To determine whether the role of STAT3 in promoting RORγt induction solely relies on c-Maf, we restored c-Maf expression in STAT3-deficient iTreg17 cells. WT and c-Maf KO CD4 T cells were infected with a control-GFP or a c-Maf-GFP encoding retrovirus. Although the c-Maf encoded retrovirus did restore RORγt expression in c-Maf-deficient Tregs and induced optimal levels of c-Maf in STAT3 KO Tregs, it was unable to restore RORγt expression in the latter cells (**Figure 6**). Collectively, these data indicate that in Tregs, an additional STAT3-driven but cMaf-independent pathway is required to promote optimal RORγt expression.

A pro-Th1 Inflammatory Environment Dampens RORγt Expression in Tregs

While IL-6 promoted RORγt expression in wild type iTregs, it surprisingly led to a marked reduction of this transcription factor in STAT3-deficient iTregs (**Figures 5F,G**), thus revealing the presence of an inhibitory pathway regulating RORγt expression. Studies by Costa-Pereira et al. have shown that in STAT3 KO fibroblasts, IL-6 signals through STAT1 and has IFNγ-like effects (26). We observed that in contrast to their wild type counterparts, STAT3 KO CD4⁺ T cells cultured in the presence of IL-6 expressed higher levels of phospho-STAT1 (**Figure S7**). This prompted us to investigate the consequences of a STAT1/Th1 inflammatory pathway on the differentiation of RORγt⁺ Tregs.

Infection with *Toxoplasma gondii* results in a highly Th1-polarized microenvironment leading to altered Treg cell homeostasis (27). In particular, the strong Th1 environment triggered by *T. gondii* infection has been shown to induce T-bet and IFN-γ expression in Treg cells (27, 28). Analysis of Tregs from the small intestine lamina propria of *T. gondii* infected C57BL/6 mice confirmed the emergence of a T-bet⁺ Treg subset (**Figures 7A,B**, upper panels). Interestingly, *T. gondii* infection resulted in a reduced frequency of tissue-associated RORγt⁺ Tregs, despite having minimal effect on Maf expression (**Figures 7A,B**, middle and lower panels). Th17 cell differentiation was also suppressed during *T. gondii*

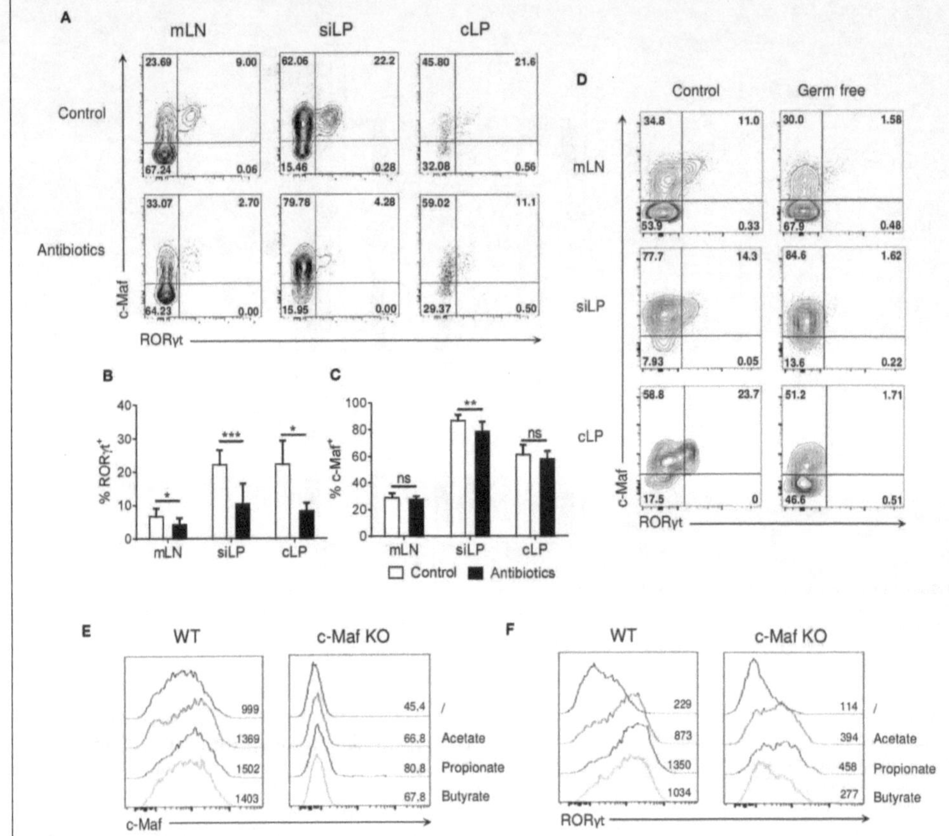

FIGURE 4 | Microbial signals induce RORγt expression in Tregs but marginally affect c-Maf expression. **(A–C)** WT mice were treated with wide-spectrum antibiotics or a control solution for 4 weeks. Representative flow cytometry expression profiles of c-Maf vs. RORγt in Treg cells **(A)** and histograms showing the frequency of RORγt+ **(B)**, and c-Maf+ **(C)** Tregs in the indicated organs (gate CD4+ Foxp3+). **(D)** Representative flow cytometry expression profiles of c-Maf vs. RORγt by Treg cells in the indicated organs of germ-free and control mice, **(E,F)** Naive WT or c-Maf-deficient CD4 T cells were activated *in vitro* for 72 h in presence of TGF-and small chain fatty acids. Histograms show the expression profiles of c-Maf **(E)** and RORγt **(F)** among Treg cells. Results are representative of at least three independent experiments; histograms in **(B,C)** represent the mean ± SD of at least four individual mice. Difference between groups is determined by an unpaired *t* test (mLN, siLP) or a Mann-Whitney test for two-tailed data (cLP), *$p < 0.05$; **$p < 0.01$; ***$p < 0.001$.

infection (**Figure S8**), consistent with the observation that Th17 cell differentiation and Th1 cell differentiation are mutually suppressive (29, 30). This reciprocal exclusion was also reflected among Tregs, as T-bet and RORγt were expressed in distinct Treg cell subsets in the infected mice (**Figure 7C**).

We next wished to evaluate the effect of Th1-promoting signals on RORγt expression by Tregs, Addition of IFN-γ to the iTreg17 differentiation media led to the selective accumulation of the phosphorylated form of STAT1, without affecting neither phospho-STAT3 accumulation nor c-Maf

expression by iTregs (**Figure S9** and **Figure 7D**, upper middle panels). Of note, IFN-γ led to the accumulation of Tbet-expressing Tregs, with a concomitant reduction in the proportions of RORγt+ Tregs (**Figure 7D**, upper left and right panels). STAT1 KO Tregs were insensitive to the IFN-γ-driven inhibition of RORγt expression (**Figure 7D**, lower panels). Altogether these results strongly suggest that the IFN-γ/STAT1 signaling pathway negatively regulates RORγt expression even in the presence of an active IL-6/STAT3-driven pathway.

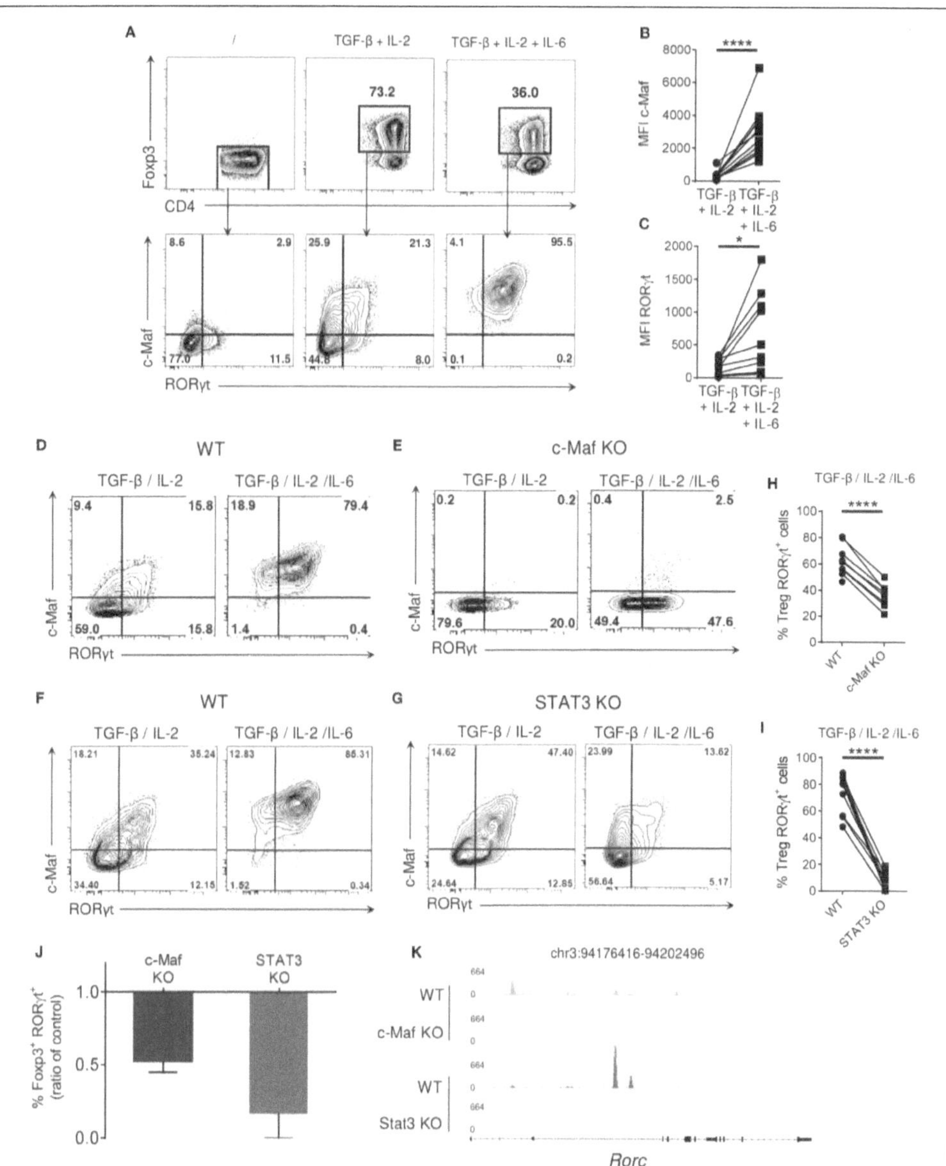

FIGURE 5 | c-Maf and STAT3 are required for optimal RORγt expression in *in vitro* polarized Tregs. WT, c-Maf or STAT3-deficient naïve CD4 T cells were activated *in vitro* in presence of polarizing cytokines for 72 h. Protein expression was then assessed by flow cytometry. **(A)** Representative flow cytometry expression profiles of Foxp3 vs. CD4 and c-Maf vs. RORγt in the indicated gate by WT Treg cells polarized *in vitro* in the indicated conditions. **(B,C)** Histograms show the MFI of c-Maf **(B)**

(Continued)

FIGURE 5 | or RORγt **(C)** in WT Treg cells *in vitro*. **(D–G)** Representative flow cytometry expression profiles of c-Maf vs. RORγt by WT **(D,F)**, c-Maf-deficient **(E)** and STAT3-deficient **(G)** Treg cells polarized in presence of TGF-and IL-2 or TGF-, IL-2, and IL-6. **(H,I)** Graphs show the frequency of RORγt+ cells among WT, c-Maf **(H)** and STAT3-deficient **(I)** Tregs polarized in presence of TGF-, IL-2, and IL-6. **(J)** Proportions of RORγt− expressing cells among CD4+ Foxp3+ Tregs is expressed as a ratio of control. Values from c-Maf or STAT3-KO Tregs were divided by the value from WT Tregs in each experimental data set. **(K)** Profiles generated from c-Maf and STAT3 ChiP-seq in WT, c-Maf- and STAT3-KO *in vitro* Th17 cells. Representative IGV tracks showing c-Maf (green), STAT3 (red) binding sites highlighted in gray at the *Rorc* locus among the indicated cell population. Gene location is indicated at the top of the panel. Y axis indicates the normalized read coverage for each track. Results are representative of at least three independent experiments. Symbols in histograms represent individual mice. Difference between groups is determined by a paired *t* test. *p < 0.05; ****p < 0.0001.

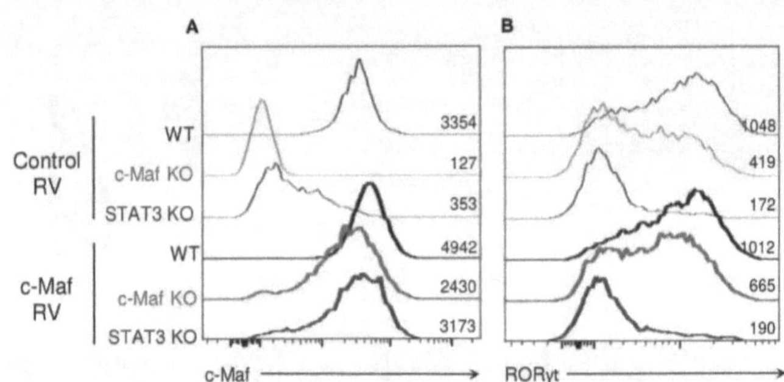

FIGURE 6 | c-Maf does not rescue RORγt expression in Tregs in the absence of STAT3. WT, c-Maf or STAT3-deficient naïve CD4 T cells were activated *in vitro* in presence of TGF-β, IL-2, and IL-6 and infected with a control or c-Maf- expressing retrovirus. c-Maf **(A)** and RORγt **(B)** expression profiles and MFI as assessed by flow cytometry after 72 h (gate CD4+ Foxp3+). Results are representative of three independent experiments.

DISCUSSION

RORγt+ Tregs form a distinct population of regulatory T cells that is crucial to maintain gastrointestinal homeostasis and prevent colitis (4, 5, 7). Whereas, RORγt+ Treg function has been amply documented, the factors driving RORγt+ Treg differentiation remain ill-defined. We show herein that multiple signaling pathways cooperate to exert both positive and negative control over RORγt+ expression by Tregs.

We confirmed and extended previous data showing that transcription factor c-Maf plays a major role for the acquisition of RORγt expression by Tregs (8, 12, 31). However, contrary to RORγt, which is confined to a subset of intestinal Tregs, c-Maf expression is found in a wider proportion of Tregs, located in distinct organs. This observation suggests that c-Maf cooperates with other pathways to induce optimal RORγt expression by these regulatory cells.

We showed that microbial signals, as well as IL-6-, STAT3-, and TGF-β-signaling pathways promote RORγt+ Treg cell differentiation in a non-redundant manner. Constrastingly, blocking any one of these pathways only had minimal effect on c-Maf expression *in vivo*, suggesting that several redundant pathways cooperate to induce c-Maf expression in Tregs.

Neumann et al. reported that, in addition to the loss of RORγt expression in Tregs, which we also observed, germ-free mice exhibited a near complete loss of c-Maf expression in intestinal Tregs. While we could not reproduce this observation, we observed that Tregs found in germ-free mice expressed slightly reduced levels of c-Maf. Reminding that the RORγt+ population expresses the highest levels of c-Maf among intestinal Tregs, we therefore speculate that RORγt expression in Tregs requires a high c-Maf expression threshold and that even a minimal decrease in c-Maf expression could hinder RORγt expression in microbiota-driven intestinal Tregs.

Short chain fatty acids (SCFA) produced by gut commensal microbes induce functional colonic Tregs and protect against T cell-dependent experimental colitis (24, 32). Depending on the cytokine environment and immunological context, butyrate, acetate, and propionate, the most available SCFA in the gut, can also support IL-10 expression in Th1 and Th17 effector cells, thereby inhibiting colitis caused by pathogenic T cells (33, 34). We show herein that SCFA induce RORγt expression in *in vitro* differentiated Tregs in a c-Maf-dependent and independent manner. Although SCFA promote peripheral Treg cell generation (24), no significant correlation was observed between any single SCFA and RORγt+ Treg frequency. Moreover, oral supplementation with a mixture of acetate, propionate, and butyrate failed to increase the frequency of mesenteric lymph node RORγt+ Tregs (4, 23). Thus, although our *in vitro* data showed that SCFA can promote RORγt expression in a pro-Treg culture medium (i.e., in the presence of TGF-β), SCFA alone cannot account for

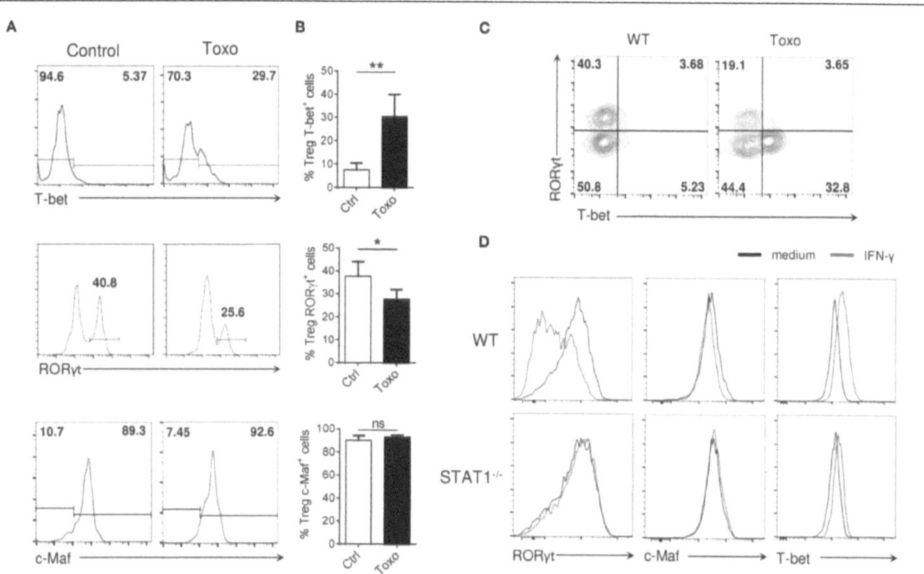

FIGURE 7 | Inflammatory Th1 responses opposes RORγt expression in Tregs. **(A)** Histograms show RORγt, c-Maf and T-bet expression among Tregs in the siLP of WT mice 8 days after infection with *Toxoplasma gondii* or control mice (gate CD4⁺ Foxp3⁺). **(B)** Frequency of RORγt⁺, c-Maf⁺, and T-bet⁺ cells among Treg cells in the siLP of WT mice 8 days after infection with *Toxoplasma gondii* or control mice. **(C)** Representative flow cytometry expression profiles of RORγt vs. T-bet among Treg cells isolated from mLN of WT mice infected or not with *Toxoplasma gondii* (gate CD4⁺ Foxp3⁺). **(D)** Histograms show RORγt, c-Maf, and T-bet expression in WT and STAT1-deficient *in vitro* Treg17 cells stimulated with or without IFN-γ (100 ng/ml) for 72 h. Results are representative of at least two **(A–C)** or three **(D)** independent experiments; histograms represent the mean ± SD of five individual mice. Difference between groups is determined by a Mann-Whitney test for two-tailed data. *$p < 0.05$.

the impact of the microbiota on intestinal RORγt⁺ Treg development *in vivo*.

The proportion of RORγt⁺ Tregs was severely reduced in mice deficient for STAT3 or TGF-β signaling. Rather surprisingly, while STAT3 and TGF-β signaling drive c-Maf expression in Tfh and Th17 cells (25, 35, 36), we found that mice deficient for STAT3 or TGF-β in the T cell compartment harbored normal to even slightly increased numbers of c-Maf⁺ Tregs, indicating that, while STAT3 and TFG-β seem dispensable for the induction of c-Maf expression, they are both essential to achieve optimal differentiation of RORγt⁺ Tregs in the intestine.

Besides c-Maf, many genes involved in Th17 differentiation, such as *Irf4, Batf, Rora, Ahr, Sox5t,* and *Hif-1α*, are expressed in response to STAT3 activation (19, 37–41). Sox5t is a T cell isoform of Sox5 which induces RORγt expression in Th17 cells via physical interaction with c-Maf (19). As enforced expression of c-Maf was not sufficient to induce Foxp3⁺ RORγt⁺ T cell differentiation in absence of STAT3, we speculate that Sox5, or other molecules downstream of STAT3, act together with c-Maf to achieve RORγt expression in Tregs. Interestingly, ChIP-seq data revealed that STAT3 and c-Maf bind with different

intensities to distinct sites of the *Rorc* locus. The inability of c-Maf to compensate STAT3-deficiency in RORγt⁺ Treg differentiation could therefore also be explained by a direct effect of STAT3 on RORγt expression. This is in agreement with previous data showing that STAT3 binds to intron 1 of *Rorc* gene and induces chromatin remodeling of the locus (39). Overall, it would seem that STAT3 controls RORγt⁺ Treg cell fate both through direct activation of the *Rorc* locus and by regulating the expression of a set of genes, including c-Maf, that is essential for RORγt⁺ Treg differentiation (**Figure S10**).

In T cells, IL-6 predominantly signals via STAT3 and to a lesser extent via STAT1. It has been proposed that the accessibility of different STATs within the cell influences the outcome of cytokine signaling (42), as illustrated by the observation that IL-6 acquired the ability to induce the expression of STAT1-dependent genes in STAT3-deficient cells (26, 43). However, Hirahara et al. showed an asymmetric action of STAT3 and STAT1 at the genomic level where much of STAT1 chromatin binding was STAT3-dependent. This challenged the classical view that, in the absence of its major STAT module, a cytokine would acquire an alternative STAT-signaling profile (44). Yet, in the absence of STAT3, and despite a global reduction in STAT1 chromatin binding, some preferential STAT1 binding sites were conserved in a group of IL-6 downregulated genes. With this in mind we hypothesized a negative influence of STAT1 on RORγt expression by Tregs. Indeed, STAT3-deficient T cells showed increased STAT1 phosphorylation in response to IL-6, compatible with a switch from STAT3 to STAT1 signaling in these cells. We further showed that IFN-γ-driven activation of STAT1 opposed RORγt expression in Tregs, without affecting Foxp3 and c-Maf expression or STAT3 phosphorylation. Naïve STAT1 KO mice did not show altered proportions of intestinal RORγt$^+$ Tregs (data not shown), which could be explained by the lack of inflammatory Th1 components at steady state. Indeed, in wild type mice infected with the Th1-prototypic *Toxoplasma gondii* intestinal parasite, RORγt expression was decreased in intestinal Tregs, confirming the antagonistic role of inflammatory Th1 responses on RORγt expression in Tregs *in vivo*.

Different integrative pathways have been proposed to explain the functional outcome of multiple STAT signaling in distinct T cell subsets (45). Meyer Zu Horste et al. recently reported that death receptor Fas promotes Th17 cell differentiation and inhibits Th1 cell development by preventing STAT1 activation. In this model, Fas regulated the STAT1 vs. STAT3 balance by binding and sequestering STAT1 (46). Although not formerly excluded, sequestration of STAT3 is unlikely in RORγt$^+$ Tregs as addition of IFN-γ to the iTreg17 culture media did not affect the phosphorylation status of STAT3. Our data rather suggest that Treg cell fate results from the balance of STAT1 and STAT3 driven signals. Gene expression could also conceivably be fine-tuned by the formation of STAT1/STAT3 heterodimers, as proposed for the IL-21 signaling (47). Of interest, patients with loss-of-function STAT3 mutations or with gain-of-function STAT1 mutations show similar susceptibility to fungal infections (48, 49). In the latter group, overactive STAT1 appears to limit STAT3-driven antifungal responses (49).

Further work is required to decipher whether STAT1 interacts with STAT3 or exerts an independent negative role on RORγt$^+$ Treg cell fate. As T-bet is induced in Tregs that develop during *Toxoplasma* infection or in response to IFN-γ, we can also envision that STAT1 signaling inhibits RORγt expression through T-bet blocking of Runx1-mediated transactivation of *Rorc*, as previously reported for the Th1/Th17 lineage specification (50). Regardless of the molecular mechanism, the observation that IFN-γ/STAT1 signaling pathway negatively regulates RORγt, even in the presence of an active IL-6/STAT3 pathway, suggests a dominant negative effect of STAT1 over STAT3 in these experimental conditions.

The antagonism between STAT1 and STAT3 seems to be cell type-specific or specific to a certain gene locus, as a cooperation between STAT1 and STAT3 downstream of IL-6 has been described for optimal Bcl6 induction and Tfh differentiation in response to viral infections (51).

Collectively, our data reveal that, beyond the previously established c-Maf/RORγt interplay, multiple signaling pathways cooperate to exert a tight control over RORγt expression in Tregs.

DATA AVAILABILITY STATEMENT

All datasets generated for this study are included in the article/**Supplementary Material**.

ETHICS STATEMENT

The animal study was reviewed and approved by Université Libre de Bruxelles Institutional Animal Care and Use Committee.

AUTHOR CONTRIBUTIONS

FA conceived and supervised the research program and experiments, and wrote the manuscript. HH performed the experiments, acquired, and analyzed data. SD, FV, AA, YA, HE-K, and GO performed experiments and contributed to the analysis of the data. OL conceived the research program and revised the manuscript.

FUNDING

This work was supported by the European Regional Development Fund (ERDF) and the Walloon Region (Wallonia-Biomed portfolio, 411132-957270), grant from the Fonds Jean Brachet and research credit from the National Fund for Scientific Research, FNRS, Belgium. FA was a Research Associate at the FNRS. YA was recipient of a research fellowship from the FNRS/Télévie. HH was supported by a Belgian FRIA fellowship.

SUPPLEMENTARY MATERIAL

The Supplementary Material for this article can be found online at: https://www.frontiersin.org/articles/10.3389/fimmu 2019.03007/full#supplementary-material

REFERENCES

1. Chen W, Jin W, Hardegen N, Lei KJ, Li L, Marinos N, et al. Conversion of peripheral CD4$^+$CD25$^-$ naive T cells to CD4$^+$CD25$^+$ regulatory T cells by TGF-beta induction of transcription factor Foxp3. *J Exp Med.* (2003) 198:1875–86. doi: 10.1084/jem.20030152

2. Whibley N, Tucci A, Powrie F. Regulatory T cell adaptation in the intestine and skin. *Nat Immunol.* (2019) 20:386–96. doi: 10.1038/s41590-019-0351-z

3. Panduro M, Benoist C, Mathis D. Tissue tregs. *Annu Rev Immunol.* (2016) 34:609–33. doi: 10.1146/annurev-immunol-032712-095948

4. Sefik E, Geva-Zatorsky N, Oh S, Konnikova L, Zemmour D, McGuire AM, et al. MUCOSAL immunology individual intestinal symbionts induce a distinct population of Rory$^+$ regulatory T cells. *Science.* (2015) 349:993–7. doi: 10.1126/science.aaa9420

5. Ohnmacht C, Park JH, Cording S, Wing JB, Atarashi K, Obata Y, et al. MUCOSAL immunology the microbiota regulates type 2 immunity through Roryt$^+$ T cells. *Science.* (2015) 349:989–93. doi: 10.1126/science.aac4263

6. Lochner M, Ohnmacht C, Presley L, Bruhns P, Si-Tahar M, Sawa S, et al. Microbiota-induced tertiary lymphoid tissues aggravate inflammatory disease in the absence of rorgamma T and LTi cells. *J Exp Med.* (2011) 208:125–34. doi: 10.1084/jem.20100052

7. Yang B-H, Hagemann S, Mamareli P, Lauer U, Hoffmann U, Beckstette M, et al. Foxp3$^+$ T cells expressing Roryt represent a stable regulatory T-cell effector lineage with enhanced suppressive capacity during intestinal inflammation. *Mucosal Immunol.* (2015) 205:1381–93. doi: 10.1038/mi.2015.74

8. Xu M, Pokrovskii M, Ding Y, Yi R, Au C, Harrison OJ, et al. c-Maf-dependent regulatory T cells mediate immunological tolerance to a gut pathobiont. *Nature.* (2018) 554:373–7. doi: 10.1038/nature25500

9. Kim BS, Lu H, Ichiyama K, Chen X, Zhang YB, Mistry NA, et al. Generation of Roryt(+) antigen-specific T regulatory 17 cells from Foxp3(+) precursors in autoimmunity. *Cell Rep.* (2017) 21:195–207. doi: 10.1016/j.celrep.2017.09.021

10. Komatsu N, Okamoto K, Sawa S, Nakashima T, Oh-hora M, Kodama T, et al. Pathogenic conversion of Foxp3$^+$ T cells into Th17 cells in autoimmune arthritis. *Nat Med.* (2014) 20:62–8. doi: 10.1038/nm.3432

11. Downs-Canner S, Berkey S, Delgoffe GM, Edwards RP, Curiel T, Odunsi K, et al. Suppressive IL-17a(+)Foxp3(+) and ex-Th17 IL-17a(neg)Foxp3(+) T(reg) cells are a source of tumour-associated T(reg) cells. *Nat Commun.* (2017) 8:14649. doi: 10.1038/ncomms14649

12. Wheaton JD, Yeh CH, Ciofani M. Cutting edge: c-Maf is required for regulatory T cells to adopt Roryt$^+$ and follicular phenotypes. *J Immunol.* (2017) 199:3931–6. doi: 10.4049/jimmunol.1701134

13. Imbratta C, Leblond MM, Bouzourène H, Speiser DE, Velin D, Verdeil G. Maf deficiency in T cells dysregulates T(reg) - T(h)17 balance leading to spontaneous colitis. *Sci Rep.* (2019) 9:6135. doi: 10.1038/s41598-019-42486-2

14. Hiramatsu Y, Suto A, Kashiwakuma D, Kanari H, Kagami S, Ikeda K, et al. C-maf activates the promoter and enhancer of the IL-21 gene, and TGF-beta inhibits c-maf-induced IL-21 production in CD4$^+$ T cells. *J Leukoc Biol.* (2010) 87:703–12. doi: 10.1189/jlb.0909639

15. Sahoo A, Alekseev A, Tanaka K, Obertas L, Lerman B, Haymaker C, et al. Batf is important for IL-4 expression in T follicular helper cells. *Nat Commun.* (2015) 6:7997. doi: 10.1038/ncomms8997

16. Apetoh L, Quintana FJ, Pot C, Joller N, Xiao S, Kumar D, et al. The aryl hydrocarbon receptor interacts with c-Maf to promote the differentiation of type 1 regulatory T cells induced by IL-27. *Nat Immunol.* (2010) 11:854–61. doi: 10.1038/ni.1912

17. Bauquet AT, Jin H, Paterson AM, Mitsdoerffer M, Ho IC, Sharpe AH, et al. The costimulatory molecule ICOS regulates the expression of c-Maf and IL-21 in the development of follicular T helper cells and Th-17 cells. *Nat Immunol.* (2009) 10:167–75. doi: 10.1038/ni.1690

18. Andris F, Denanglaire S, Anciaux M, Hercor M, Hussein H, Leo O. The transcription factor c-Maf promotes the differentiation of follicular helper T cells. *Front Immunol.* (2017) 8:480. doi: 10.3389/fimmu.2017.00480

19. Tanaka S, Suto A, Iwamoto T, Kashiwakuma D, Kagami S, Suzuki K, et al. Sox5 and c-Maf cooperatively induce Th17 cell differentiation via Roryt induction as downstream targets of STAT3. *J Exp Med.* (2014) 211:1857–74. doi: 10.1084/jem.20130791

20. Rubtsov YP, Rasmussen JP, Chi EY, Fontenot J, Castelli L, Ye X, et al. Regulatory T cell-derived interleukin-10 limits inflammation at environmental interfaces. *Immunity.* (2008) 28:546–58. doi: 10.1016/j.immuni.2008.02.017

21. Chytil A, Magnuson MA, Wright CV, Moses HL. Conditional inactivation of the TGF-β type Ii receptor using Cre:Lox. *Genesis.* (2002) 32:73–5. doi: 10.1002/gene.10046

22. Veldhoen M, Hocking RJ, Atkins CJ, Locksley RM, Stockinger B. TGFβ in the context of an inflammatory cytokine milieu supports *de novo* differentiation of IL-17-producing T cells. *Immunity.* (2006) 24:179–89. doi: 10.1016/j.immuni.2006.01.001

23. Abdel-Gadir A, Stephen-Victor E, Gerber GK, Noval Rivas M, Wang S, Harb H, et al. Microbiota therapy acts via a regulatory T cell MyD88/Roryt pathway to suppress food allergy. *Nat Med.* (2019) 25:1164–74. doi: 10.1038/s41591-019-0461-z

24. Arpaia N, Campbell C, Fan X, Dikiy S, van der Veeken J, deRoos P, et al. Metabolites produced by commensal bacteria promote peripheral regulatory T-cell generation. *Nature.* (2013) 504:451–5. doi: 10.1038/nature12726

25. Ciofani M, Madar A, Galan C, Sellars M, Mace K, Pauli F, et al. A validated regulatory network for Th17 cell specification. *Cell.* (2012) 151:289–303. doi: 10.1016/j.cell.2012.09.016

26. Costa-Pereira AP, Tininini S, Strobl B, Alonzi T, Schlaak JF, Is'harc H, et al. Mutational -switch of an IL-6 response to an interferon-γ-like response. *Proc Natl Acad Sci USA.* (2002) 99:8043–7. doi: 10.1073/pnas.122236099

27. Oldenhove G, Bouladoux N, Wohlfert EA, Hall JA, Chou D, Dos Santos L, et al. Decrease of Foxp3$^+$ treg cell number and acquisition of effector cell phenotype during lethal infection. *Immunity.* (2009) 31:772–86. doi: 10.1016/j.immuni.2009.10.001

28. Hall AO, Beiting DP, Tato C, John B, Oldenhove G, Lombana CG, et al. The cytokines interleukin 27 and interferon-γ promote distinct treg cell populations required to limit infection-induced pathology. *Immunity.* (2012) 37:511–23. doi: 10.1016/j.immuni.2012.06.014

29. Harrington LE, Hatton RD, Mangan PR, Turner H, Murphy TL, Murphy KM, et al. Interleukin 17-producing CD4$^+$ effector T cells develop via a lineage distinct from the T helper type 1 and 2 lineages. *Nat Immunol.* (2005) 6:1123–32. doi: 10.1038/ni1254

30. Park H, Li Z, Yang XO, Chang SH, Nurieva R, Wang YH, et al. A distinct lineage of CD4 T cells regulates tissue inflammation by producing interleukin 17. *Nat Immunol.* (2005) 6:1133–41. doi: 10.1038/ni1261

31. Neumann C, Blume J, Roy U, Teh PP, Vasanthakumar A, Beller A, et al. C-maf-dependent treg cell control of intestinal Th17 cells and iga establishes host-microbiota homeostasis. *Nat Immunol.* (2019) 20:471–81. doi: 10.1038/s41590-019-0316-2

32. Furusawa Y, Obata Y, Fukuda S, Endo TA, Nakato G, Takahashi D, et al. Commensal microbe-derived butyrate induces the differentiation of colonic regulatory T cells. *Nature.* (2013) 504:446–50. doi: 10.1038/nature12721

33. Park J, Kim M, Kang SG, Jannasch AH, Cooper B, Patterson J, et al. Short-chain fatty acids induce both effector and regulatory T cells by suppression of histone deacetylases and regulation of the mtor-s6k pathway. *Mucosal Immunol.* (2015) 8:80–93. doi: 10.1038/mi.2014.44

34. Sun M, Wu W, Chen L, Yang W, Huang X, Ma C, et al. Microbiota-derived short-chain fatty acids promote TH1 cell IL-10 production to maintain intestinal homeostasis. *Nat Commun.* (2018) 9:3555. doi: 10.1038/s41467-018-05901-2

35. Rutz S, Noubade R, Eidenschenk C, Ota N, Zeng W, Zheng Y, et al. Transcription factor c-Maf mediates the TGF-[beta]-dependent suppression of IL-22 production in Th17 cells. *Nat Immunol.* (2011) 12:1238–45. doi: 10.1038/ni.2134

36. Mari N, Hercor M, Denanglaire S, Leo O, Andris F. The capacity of Th2 lymphocytes to deliver B-cell help requires expression of the transcription factor STAT3. *Eur J Immunol.* (2013) 43:1489–98. doi: 10.1002/eji.201242938

37. Brüstle A, Heink S, Huber M, Rosenplänter C, Stadelmann C, Yu P, et al. The development of inflammatory Th-17 cells requires interferon-regulatory factor 4. *Nat Immunol.* (2007) 8:958. doi: 10.1038/ni1500

38. Dang EV, Barbi J, Yang HY, Jinasena D, Yu H, Zheng Y, et al. Control of T(h)17/T(reg) balance by hypoxia-inducible factor 1. *Cell.* (2011) 146:772–84. doi: 10.1016/j.cell.2011.07.033

39. Durant L, Watford WT, Ramos HL, Laurence A, Vahedi G, Wei L, et al. Diverse targets of the transcription factor STAT3 contribute to t cell pathogenicity and homeostasis. *Immunity.* (2010) 32:605–15. doi: 10.1016/j.immuni.2010.05.003

40. Schraml BU, Hildner K, Ise W, Lee W-L, Smith WA-E, Solomon B, et al. The AP-1 transcription factor batf controls Th17 differentiation. *Nature.* (2009) 460:405. doi: 10.1038/nature08114

41. Yang XO, Pappu BP, Nurieva R, Akimzhanov A, Kang HS, Chung Y, et al. T helper 17 lineage differentiation is programmed by orphan nuclear receptors RORα and RORγ. *Immunity.* (2008) 28:29–39. doi: 10.1016/j.immuni.2007.11.016

42. Regis G, Pensa S, Boselli D, Novelli F, Poli V. Ups and downs: the STAT1:STAT3 seesaw of interferon and gp130 receptor signalling. *Semin Cell Dev Biol.* (2008) 19:351–9. doi: 10.1016/j.semcdb.2008.06.004

43. Schiavone D, Avalle L, Dewilde S, Poli V. The immediate early genes fos and egr1 become STAT1 transcriptional targets in the absence of STAT3. *FEBS Lett.* (2011) 585:2455–60. doi: 10.1016/j.febslet.2011.06.020

44. Hirahara K, Onodera A, Villarino AV, Bonelli M, Sciumè G, Laurence A, et al. Asymmetric action of STAT transcription factors drives transcriptional outputs and cytokine specificity. *Immunity.* (2015) 42:877–89. doi: 10.1016/j.immuni.2015.04.014

45. Lin JX, Leonard WJ. Fine-tuning cytokine signals. *Annu Rev Immunol.* (2019) 37:295–324. doi: 10.1146/annurev-immunol-042718-041447

46. Meyer Zu Horste G, Przybylski D, Schramm MA, Wang C, Schnell A, Lee Y, et al. Fas promotes T helper 17 cell differentiation and inhibits T helper 1 cell development by binding and sequestering transcription factor STAT1. *Immunity.* (2018) 48:556–69.e7. doi: 10.1016/j.immuni.2018.03.008

47. Wan CK, Andraski AB, Spolski R, Li P, Kazemian M, Oh J, et al. Opposing roles of STAT1 and STAT3 in IL-21 function in CD4[+] T cells. *Proc Natl Acad Sci USA.* (2015) 112:9394–9. doi: 10.1073/pnas.1511711112

48. O'Shea JJ, Holland SM, Staudt LM. JAKs and stats in immunity, immunodeficiency, and cancer. *N Engl J Med.* (2013) 368:161–70. doi: 10.1056/NEJMra1202117

49. Casanova JL, Holland SM, Notarangelo LD. Inborn errors of human jaks and stats. *Immunity.* (2012) 36:515–28. doi: 10.1016/j.immuni.2012.03.016

50. Lazarevic V, Chen X, Shim J-H, Hwang E-S, Jang E, Bolm AN, et al. T-bet represses Th17 differentiation by preventing runx1-mediated activation of the gene encoding Rorγt. *Nat Immunol.* (2011) 12:96–104. doi: 10.1038/ni.1969

51. Choi YS, Eto D, Yang JA, Lao C, Crotty S. Cutting edge: stat1 is required for IL-6 - mediated Bcl6 induction for early follicular helper cell differentiation. *J Immunol.* (2013) 190:3049–53. doi: 10.4049/jimmunol.1203032

52. Hussein H, Denanglaire S, Van Gool F, Azouz A, Ajouaou Y, El-Khatib H, et al. Multiple environmental signaling pathways control the differentiation of RORγt-expressing regulatory T cells. *bioRxiv. [Preprint].* (2019). doi: 10.1101/791244

Conflict of Interest: The authors declare that the research was conducted in the absence of any commercial or financial relationships that could be construed as a potential conflict of interest.

Copyright © 2020 Hussein, Denanglaire, Van Gool, Azouz, Ajouaou, El-Khatib, Oldenhove, Leo and Andris. This is an open-access article distributed under the terms of the Creative Commons Attribution License (CC BY). The use, distribution or reproduction in other forums is permitted, provided the original author(s) and the copyright owner(s) are credited and that the original publication in this journal is cited, in accordance with accepted academic practice. No use, distribution or reproduction is permitted which does not comply with these terms.

Supplemental figures

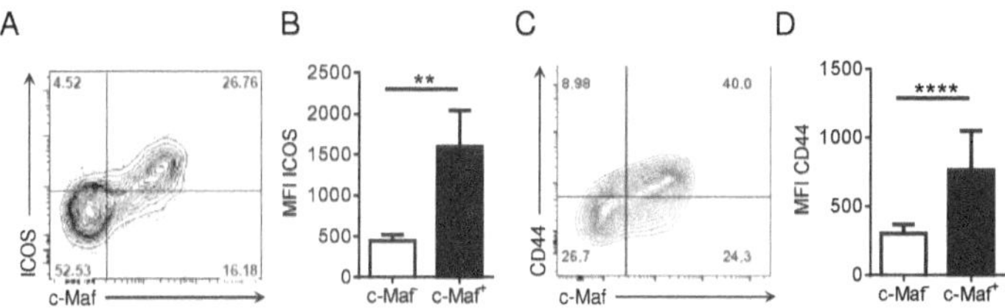

Figure S1. **c-Maf⁺ Tregs have an activated phenotype.** (A, C) Representative flow cytometry expression profiles of ICOS (A) or CD44 (C) versus c-Maf among Treg cells in mLN of WT mice (gate CD4⁺ Foxp3⁺). (B, D) Histograms show the MFI of ICOS (B) and CD44 (D) among c-Maf⁻ and c-Maf⁺ Treg cells. Histograms represent the mean ± SD of at least five individual mice. Difference between groups is determined by a Mann–Whitney test for two-tailed data. **p < 0.01; ****p < 0.0001

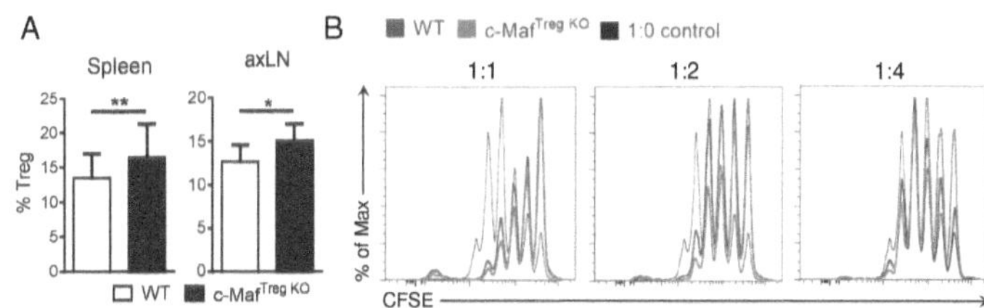

Figure S2. **c-Maf-deficient Tregs have a suppressive phenotype.** (A) Frequency of Treg cells in the spleen or axillary lymph nodes of WT and c-Maf^{Treg KO} mice (gate CD4⁺). (B) Histograms showing CFSE staining profiles of conventional CD4 T cells in a Treg suppression assay *in vitro*. Results are representative of at least three independent experiments; histograms represent the mean ± SD of five individual mice. Difference between groups is determined by a Mann–Whitney test for two-tailed data. *p < 0.05; **p < 0.01

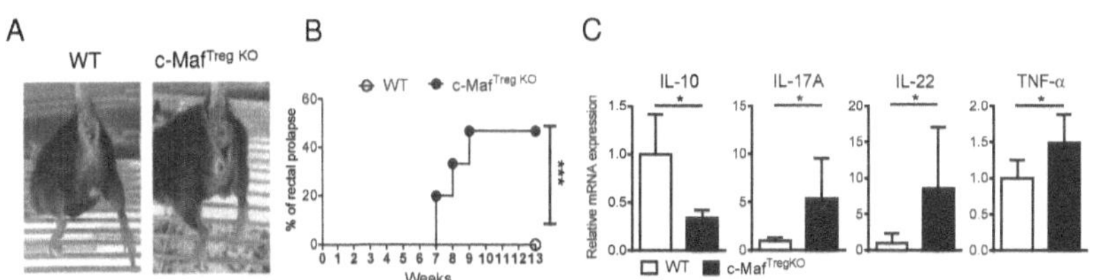

Figure S3. **c-Maf-deficient Tregs are unable to control spontaneous Th17 intestinal responses.** (A) Rectal prolapse appears in c-Maf^{Treg KO} mice. (B) Incidence of rectal prolapse in WT and c-Maf^{Treg KO} mice. (C) Histograms show mRNA expression of IL-10, IL-17A, IL-22 and TNF-α in total WT or c-Maf^{Treg KO} mLN relative to RPL32. Results are representative of at least three independent experiments; histograms represent the mean ± SD of five individual mice. Difference between groups is determined by a Mantel-Cox test (B) or a Mann–Whitney test for two-tailed data. *p < 0.05; ***p < 0.001

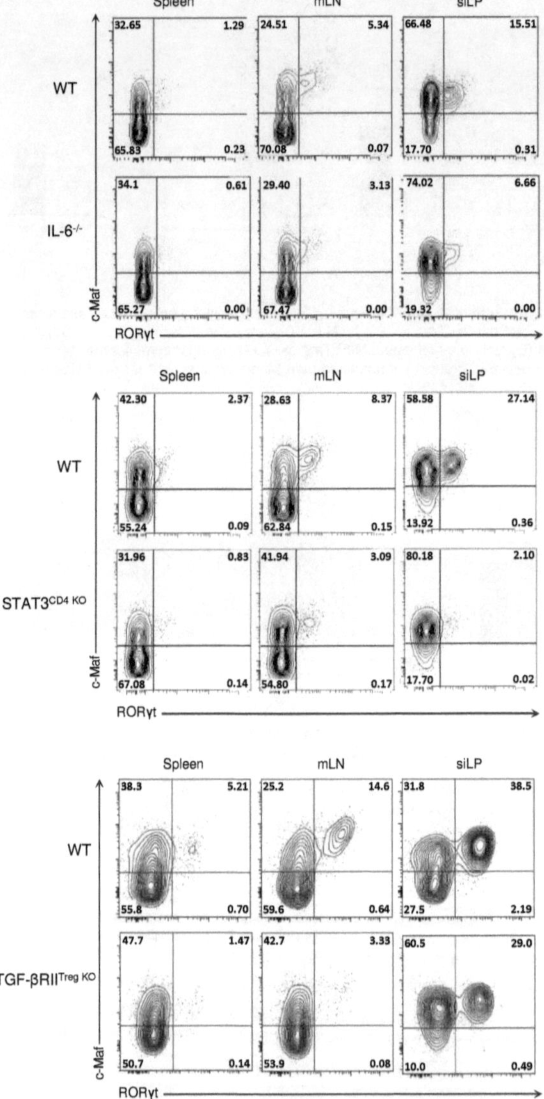

Figure S4. **IL-6/STAT3 and TGF-β signaling promote RORγt expression in Tregs independently of c-Maf.** Representative flow cytometry expression profiles of c-Maf versus RORγt in Treg cells in the indicated organs of the indicated mice strains (gate CD4+ Foxp3+). Results are representative of at least three independent experiments.

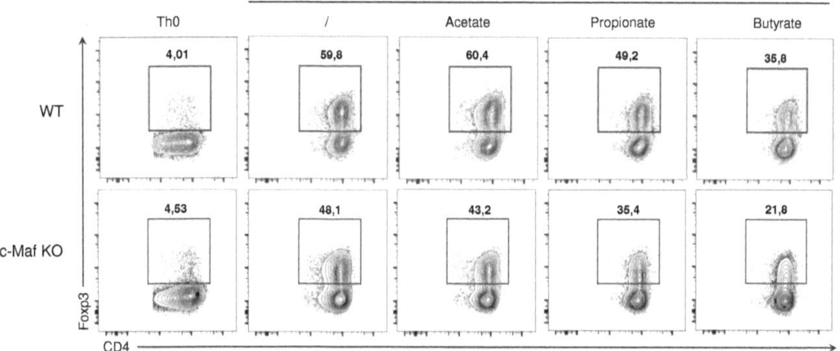

Figure S5. **In vitro Treg differentiation in presence of short-chain fatty acids.** Naïve WT or c-Maf-deficient CD4 T cells were activated *in vitro* for 72h in presence of TGF-β and small chain fatty acids. Representative flow cytometry expression profiles of Foxp3 in CD4 cells in the indicated conditions; gating strategy of Fig. 4E, F. Results are representative of at least three independent experiments.

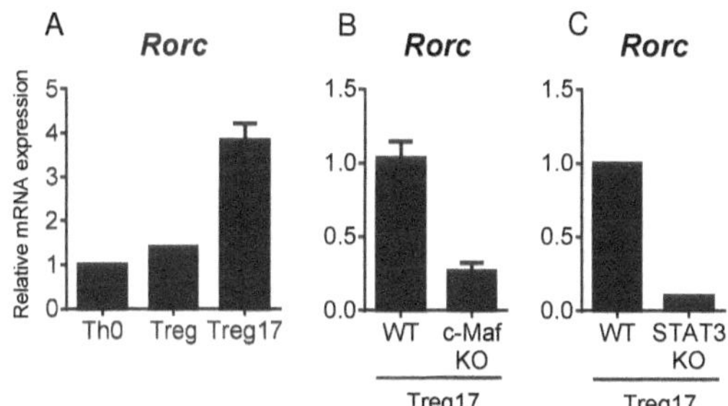

Figure S6. **Transcription control of RORγt expression in Tregs.** Naïve CD4 T cells from Foxp3-Cre-YFP (A, B), c-Maf[Treg KO] (B), or WT and STAT3 KO (C) mice were activated *in vitro* in presence of polarizing cytokines for 72h (Treg: TGF-β, IL-2; Treg17: TGF-β, IL-2, IL-6). Treg cells from A and B were further purified by FACS based on YFP expression. Histograms show mRNA expression of *Rorc* relative to RPL32, with Th0 (A) or WT Treg17 (B, C) set to 1. Results are representative of at least three independent experiments; histograms represent the mean ± SD of 2-3 individual samples.

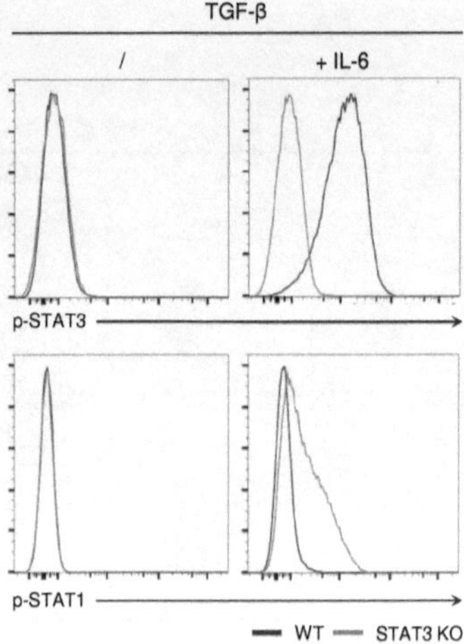

Figure S7. **In absence of STAT3 in Tregs, IL-6 signals via STAT1.** Histograms show the expression of pSTAT3 and pSTAT1 among Treg cells polarized *in vitro* in presence of TGF-β, with or without IL-6 (gate CD4+).

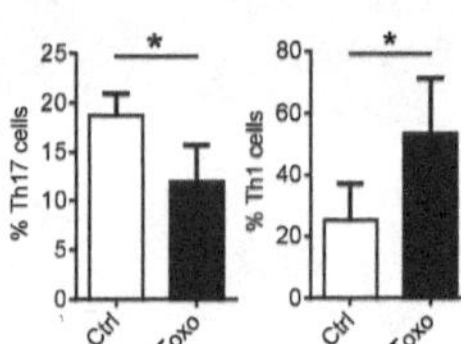

Figure S8. **Infection with *Toxoplasma gondii* is characterized by a Th1 inflammatory environment.** Histograms show the frequency of Th17 cells and Th1 cells in siLP of mice after *Toxoplasma gondii* infection. Histograms represent the mean ± SD of five individual mice. Difference between groups is determined by a Mann–Whitney test for two-tailed data. **p < 0.01; ****p < 0.0001

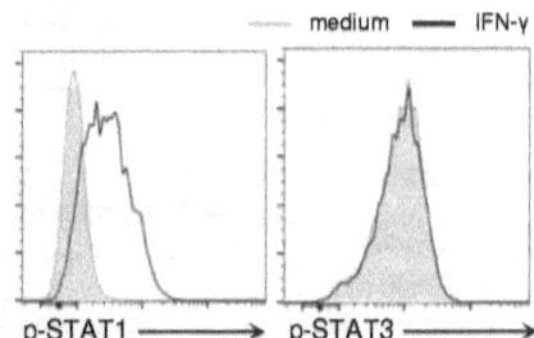

Figure S9. **IFN-γ does not impede IL-6-mediated STAT3 phosphorylation in iTregs.** Histograms show the expression of pSTAT1 and pSTAT3 among Treg cells polarized *in vitro* in presence of TGF-β, IL-2 and IL-6 with or without IFN-γ (10 ng/mL, gate CD4+ Foxp3+).

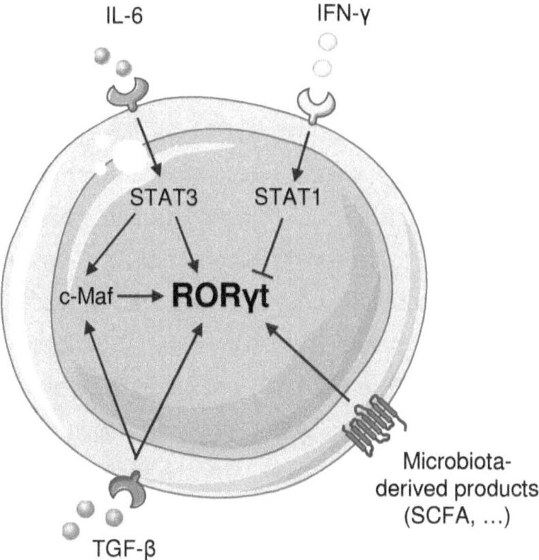

Figure S10. **Model of RORγt induction in Tregs.** The differentiation of specialized cell subsets is a balancing act: multiple positive and negative signals are integrated in order to tailor cell specialization to the immune context. Signals derived from a complex microbiota, or IL-6/STAT3 and TGF-β signaling induce RORγt expression in Tregs in a c-Maf-dependent or independent fashion. Contrastingly, inflammatory IFN-γ/STAT1 signaling opposes RORγt expression in Tregs in a c-Maf independent fashion.

3.2.2. Complementary results

3.2.2.1. IL-27 represses RORγt expression in Treg cells in a STAT1-dependent manner

By studying the signals responsible for the induction of RORγt expression in Treg cells, we have shown that the control of RORγt[+] Treg development depends on positive as well as negative signals. We have notably shown that transcription factors STAT3 and STAT1 played key antagonistic roles in this process, by respectively promoting and opposing RORγt expression in Treg cells. Interestingly, IL-27, a cytokine involved in type 1 inflammation, relies on both STAT3 and STAT1 for its signaling[255]. Additionally, the differentiation of T-bet[+] Treg cells, a subset required for the control of inflammation during *Toxoplasma gondii* infection, relies on IL-27 in the intestine, while it relies on IFN-γ in the periphery[271]. We first wanted to determine the effect of IL-27 on STAT signaling in Treg cells. Using a previously described model of *in vitro* Treg polarization, we assessed the respective activation of STAT1 and STAT3 by IL-27. Classically polarized iTreg cells (TGF-β + IL-2) show low STAT1 and STAT3 phosphorylation at 72 hours (**Figure 39A**). Expectedly, IL-6 signaling induces a strong phosphorylation of STAT3 in iTreg cells, without inducing STAT1 phosphorylation. On the other hand, IL-27 induces the phosphorylation of both STAT1 and STAT3, but induces a weaker STAT3 phosphorylation compared to IL-6. STAT1 deficiency prevents IL-27 from inducing STAT1 phosphorylation, but leads to an increase in STAT3 phosphorylation, supporting an antagonism between STAT1 and STAT3 signaling.

We next wanted to determine the role of IL-27 in regulating the phenotype of Treg cells, particularly its role on the expression of RORγt. Classically polarized iTreg cells express low levels of c-Maf and RORγt expression (**Figure 39B**), with less than 20% of the cells expressing both c-Maf and RORγt. Addition of IL-6 induces the expression of c-Maf in over 80% of Treg cells, and concomitantly induces RORγt in about 40% of Treg cells, in a STAT1-independent manner. While IL-27 induces the expression of c-Maf in over 50% of Treg cells, it represses RORγt expression to lower levels than those observed in the classical Treg polarizing condition. Indeed, fewer than 5% of Treg cells express RORγt. This repressive effect over RORγt expression is lost in absence of STAT1, but the positive

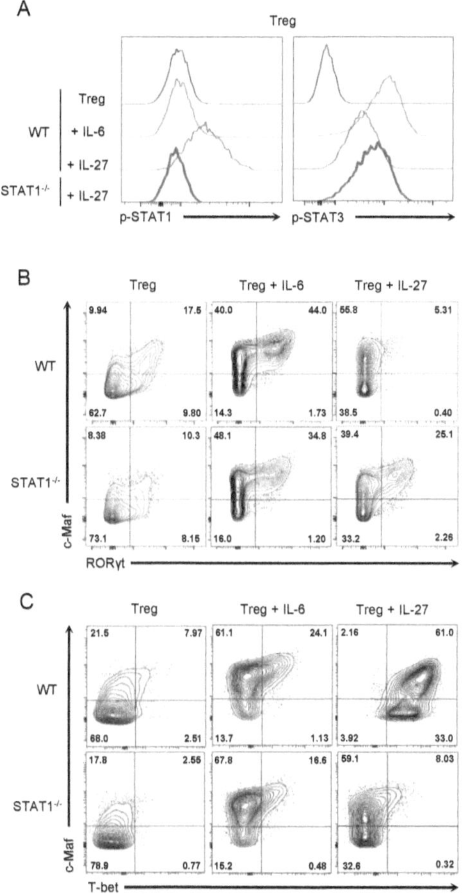

Figure 39. **IL-27 antagonizes RORγt expression in Treg cells in a STAT1-dependent manner.** Representative flow cytometry expression profiles of pSTAT1 and pSTAT3 (A; gate CD4), c-Maf versus RORγt (B; gate CD4+ Foxp3+), and c-Maf versus T-bet (C; gate CD4+ Foxp3+) among WT and STAT1-deficient Treg cells polarized *in vitro* in the indicated conditions. Results are representative of at least three independent experiments. (Treg: TGF-β, IL-2).

effect over c-Maf expression is not affected. While addition of IL-6 induces a modest amount of T-bet$^+$ Treg cells (**Figure 39C**), IL-27 induces T-bet in over 90% of Treg cells, in a highly STAT1-dependent manner. In conclusion, IL-27 promotes a switch in the Treg phenotype, by suppressing RORγt expression and concomitantly favoring T-bet expression in Treg cells, in a STAT1-depependent manner. Whether IL-27 promotes the decline of RORγt$^+$ Treg cells and the *de novo* differentiation of T-bet$^+$ Treg cells, or whether it promotes a switch between RORγt and T-bet expression in the existing pool of Treg cells remains to be defined.

3.2.2.2. c-Maf deficiency hinders in vitro Treg differentiation in presence of IL-6

Taking advantage of our model of *in vitro* Treg polarization, we observed that c-Maf deficiency does not alter the frequency of Treg cells in classical Treg polarizing conditions, over 90% of both WT and c-Maf-deficient cells differentiating into Foxp3$^+$ cells (**Figures 40A, 40B**). Only about 70% of cells expressed Foxp3 after addition of IL-6. Indeed, IL-6 is known to destabilize Foxp3 expression in T cells, and thus hinder Treg cell differentiation[431]. In absence of c-Maf, this effect was exacerbated, leading to only about 40% of cells expressing Foxp3 after addition of IL-6. Similarly to our own results (**Figures 31A, 31B**), excessive IL-6 production was associated with increased Treg frequencies and decreased Foxp3 expression among Treg cells[432]. This suggests that c-Maf could be required for the maintenance of the Treg phenotype in strong inflammatory conditions, which might be instrumental in chronic models of strong inflammation, such as the AOM/DSS model.

3.2.2.3. Transient expression of c-Maf and RORγt in Treg cells is inhibited by IL-2 and stabilized by IL-6

TGF-β promotes high c-Maf and RORγt expression in about 50% of Treg cells after three days of activation (**Figures 41A, 41B**). While the addition of IL-2 promotes Treg differentiation, it leads to decreased frequencies of c-Maf$^+$ and RORγt$^+$ cells among Treg cells, suggesting that IL-2 plays an inhibitory role over c-Maf and RORγt expression in Treg cells (**Figures 41C, 41D**). On the contrary, addition of IL-6, while decreasing Treg differentiation, promotes c-Maf and RORγt expression, leading to over 80% of c-Maf$^+$ RORγt$^+$ Treg cells in the Tr17 condition.

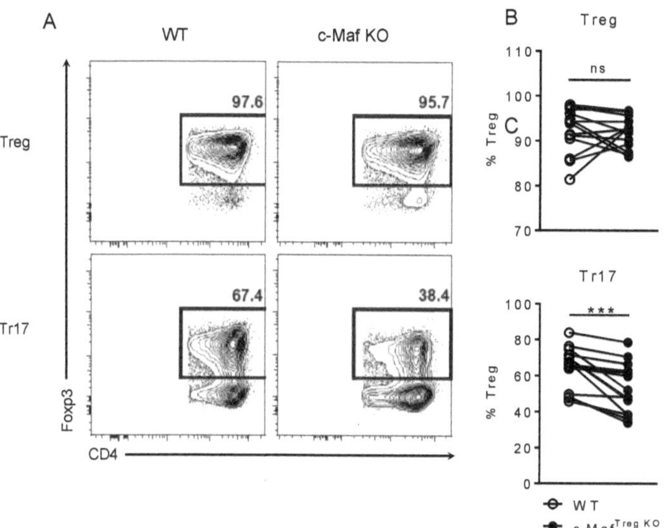

Figure 40. **c-Maf deficiency hinders *in vitro* Treg differentiation in presence of IL-6.** WT and CD4 KO naïve CD4 T cells were activated *in vitro* in presence of polarizing cytokines for 72 hours. (A) Representative flow cytometry expression profiles of Foxp3 versus CD4 (gate: live cells). (B, C) Individuals dots show the frequency of Treg in WT and c-Maf KO Treg cells polarized in Treg (B) or Tr17 (C) conditions. Results are representative of at least ten independent experiments. Histograms represent the mean ± SD. Difference between groups is determined by a paired t test. ns: non-significant; ***p < 0.001. (Treg: TGF-β, IL-2; Tr17: TGF-β, IL-2, IL-6).

We then wanted to determine the kinetics involved in the differentiation of c-Maf[+] and RORγt[+] Treg cells. Interestingly, while Treg cells polarized in presence of TGF-β and IL-2 express low levels of c-Maf and RORγt at 72 hours, over 70% of these cells are c-Maf[+] RORγt[+] at 24 and 48 hours (**Figure 41E**), indicating an early and transient expression of c-Maf and RORγt in Treg cells *in vitro*. A similar pattern of expression is observed in the Tr17 condition at 24 hours. Unlike the classical Treg condition, however, most Treg cells in the Tr17 condition maintain a stable expression of c-Maf and RORγt until day 3. These results show that c-Maf and RORγt are transiently expressed by Treg cells at an early timing. Contrary to the inhibitory effect of IL-2, IL-6 stabilizes c-Maf and RORγt expression and strongly promotes the differentiation of c-Maf[+] RORγt[+] Treg cells. While STAT5 is activated in both the Treg and Tr17 condition, STAT3 is only activated in the Tr17 condition (**Figure 41F**), suggesting that the stabilizing role of IL-6 over c-Maf and RORγt expression in Treg cells could depend on STAT3. The loss of c-Maf and RORγt expression in the Treg condition would therefore be attributed to the lack of STAT3 phosphorylation. Alternatively, these results could illustrate the inhibitory effect of IL-2 over c-Maf and RORγt expression in Treg cells.

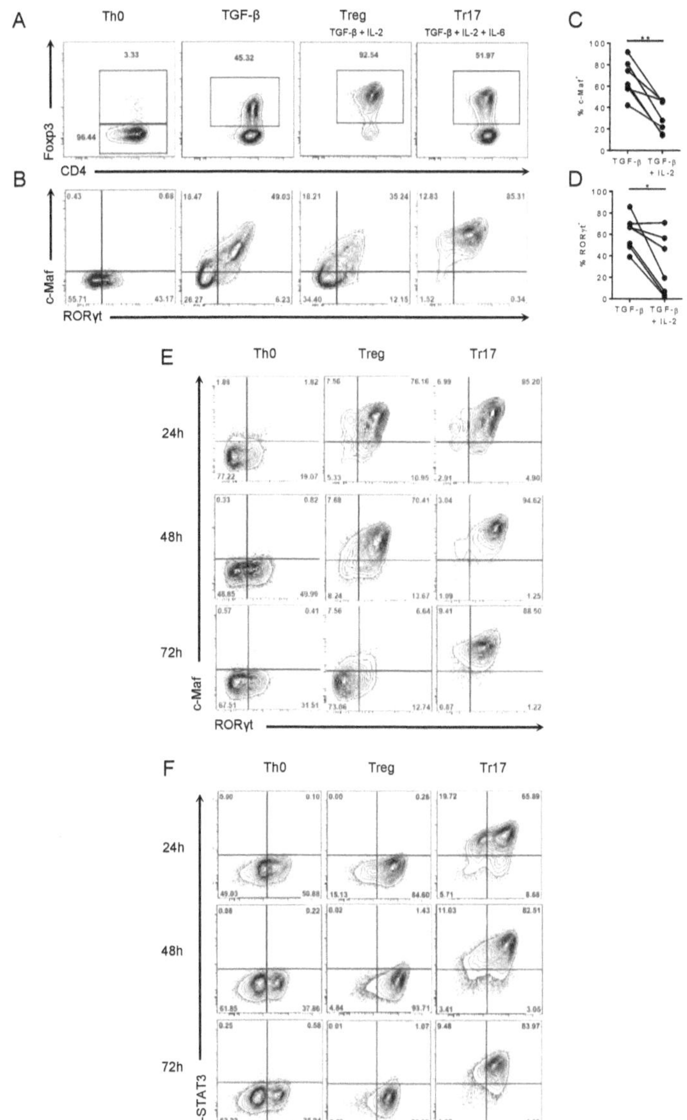

Figure 41. **Transient c-Maf and RORγt expression during *in vitro* Treg polarization.** WT naïve CD4 T cells were activated *in vitro* in presence of polarizing cytokines for 72 hours (A-D), or 24, 48, and 72 hours (E, F). Representative flow cytometry expression profiles of Foxp3 versus CD4 (A; gate: live cells), c-Maf versus RORγt (B; gate: CD4⁺ Foxp3⁺), c-Maf versus RORγt (E; gate Foxp3⁺), and p-STAT3 versus p-STAT5 (F; gate CD4⁺) in the indicated conditions. (C, D) Individuals dots show the frequency of c-Maf⁺ (C) or RORγt⁺ (D) cells among Treg cells polarized in presence of TGF-β or in Treg conditions. Histograms represent the mean ± SD. Difference between groups is determined by a paired t test. *p < 0.05; **p < 0.01. (Th0: no polarizing cytokines; Treg: TGF-β, IL-2; Tr17: TGF-β, IL-2, IL-6).

Chapter 4: General discussion and perspectives

4.1. c-Maf controls intestinal Treg specialization

During this work, we investigated the role of transcription factor c-Maf in the differentiation and function of Treg cells. We have shown that c-Maf is not required for the general homeostasis of Treg cells. Indeed, c-Maf deletion does not prevent Treg differentiation or activation[305,309,433], nor does it grossly impact Treg suppressive capacity, as evidenced by the ability of c-Maf-deficient Treg cells to suppress T cell proliferation *in vitro* and the absence of systemic inflammation symptoms in c-Maf[Treg KO] mice. While c-Maf can generally be considered as a marker of Treg activation, we have also identified a role for c-Maf in the control of intestinal Treg specialization (summarized in **Figure 42**).

It has now become clear that, while they are responsible for the general maintenance of tolerance throughout the body, Treg cells exhibit tissue-specific functions[434]. In the intestine, Treg cells act as switchmen, which discriminate between beneficial and pathogenic microbial species, thus promoting the maintenance of a healthy microbiota[435]. A disrupted tolerance to the microbiota leads to dysbiosis, which comes with multiple adverse effects, that can manifest locally, like in inflammatory bowel disease, but also systemically. Tissue-specific functions are conducted by specialized Treg subsets, which adapt their phenotype in response to environmental signals. RORγt⁺ Treg cells have been shown to control immune responses to the microbiota[269,303,307]. They arise in response to a broad, but specific, array of microbial species[269,303], in a TCR-dependent manner[307]. Of note, while commensal segmented filamentous bacteria (SFB) induced Th17 cell differentiation, *Helicobacter hepaticus* preferentially induced tolerogenic RORγt⁺ Treg cells[307], indicating that this subset is likely induced in response to pathobionts.

The preferential expression of c-Maf in intestinal Treg cells prompted several teams to investigate the role of this transcription factor in the control of gut homeostasis[305,307–309,433]. We have shown that c-Maf endows Treg cells with the ability to control homeostatic Th17 immune responses in the intestine[305]. Indeed, the specific deletion of c-Maf in Treg cells leads to the spontaneous development of a low-grade colitis in mice, characterized by the frequent apparition of a rectal prolapse, accompanied with an exacerbated Th17 response

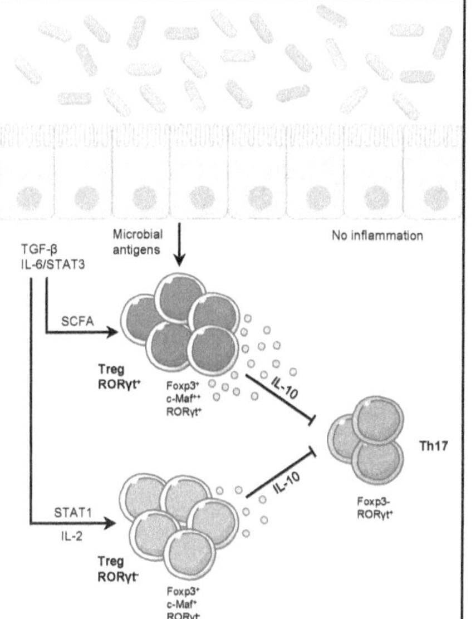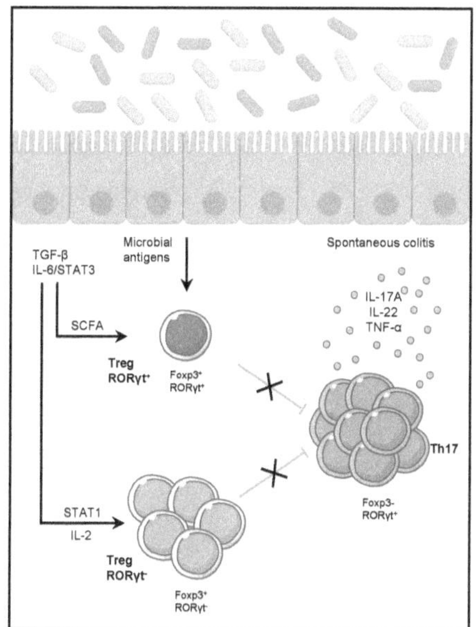

Figure 42. **c-Maf⁺ Treg cells control homeostatic Th17 intestinal inflammation.** Intestinal Treg cells acquire the expression of c-Maf in response to various local signals (antigen activation, co-stimulatory signals, TGF-β, IL-6/STAT3, SCFA). c-Maf induces RORγt expression in the RORγt⁺ Treg population, but other signals possibly constrain RORγt expression in a proportion of c-Maf⁺ Treg cells. c-Maf⁺ Treg cells control Th17 intestinal inflammation by acquiring the RORγt⁺ phenotype and by producing the anti-inflammatory cytokine IL-10. In absence of c-Maf in Treg cells (c-Maf^Treg KO), RORγt⁺ Treg cells and IL-10 expression among Treg cells are lost. Th17 cell responses are no longer controled, leading to an increased production of pro-inflammatory cytokines (IL-17A, IL-22, TNF-α) and the development of colitis.

and decreased levels of IL-10[433]. Interestingly, while the control exerted by c-Maf over gut homeostasis has been made clear by different groups, the extent of that control remains controversial.

We and others have firmly established the role of c-Maf in the differentiation of RORγt[+] Treg cells[305,307–309,433]. RORγt[+] Treg cells express high levels of immunomodulatory molecules and show enhanced suppressive abilities[269,303]. While the specificity of RORγt[+] Treg cell suppression is controversial, RORγt[+] Treg cells seem predisposed to suppress Th17 cells. Indeed, RORγt[+] Treg cells express high levels of IL-10[303,305,309], a cytokine required for the efficient suppression of Th17 cells, and also express chemokine receptor CCR6[303], which allows them to colocalize with Th17 cells. However, while RORγt[Treg KO] (*Rorc*[fl/fl] x *Foxp3*[YFPCre]) mice show increased sensitivity to different models of intestinal inflammation[268,269], they do not develop spontaneous colitis[307,309], contrary to c-Maf[Treg KO] mice[307,308,433]. Moreover, RORγt[Treg KO] mice seem able to control spontaneous Th17 intestinal responses, as they do not exhibit increased IL-17A or IL-22 levels, nor increased Th17 cell frequencies in the intestine at steady state[309]. This indicates that the ability of Treg cells to control Th17 immunity depends more strongly on c-Maf than RORγt, highlighting a role for c-Maf in Treg cells beyond the induction of RORγt expression.

We found that c-Maf deletion leads to a decreased production of IL-10 by Treg cells. Wheaton et al. argue that this decrease in IL-10 after c-Maf deletion reflects the loss of RORγt[+] Treg cells, the main IL-10-producing Treg subset[305]. We however show that IL-10 is decreased in intestinal Treg cells, irrespective of RORγt expression. Neumann et al. additionally show that IL-10 is decreased in both Helios[+] and Helios[−] intestinal Treg subsets, which supports the hypothesis that c-Maf controls IL-10 expression in all intestinal Treg cells. Neumann et al. further show that c-Maf is co-expressed with Blimp-1, IRF4, and AhR, three transcription factors that have previously been involved in the control of IL-10 expression in Treg cells or other Th subsets[309]. c-Maf inactivation in Treg cells does not majorly affect their respective expression, suggesting that the loss of IL-10 expression in c-Maf[Treg KO] mice directly depends on c-Maf. The production of IL-10 by intestinal Treg cells is essential for the preservation of gut homeostasis and IL-10 deletion in Treg cells leads to the development of spontaneous colitis[217], similar to what is observed in c-Maf$^{Treg\ KO}$ mice. It is therefore highly likely that the phenotype observed in c-Maf$^{Treg\ KO}$ mice strongly depends on the role of c-Maf in the control of IL-10 production by Treg cells.

Interestingly, Neumann et al. showed that c-Maf$^{Treg\ KO}$ mice exhibit an exacerbated IgA response, with higher IgA coating of the gut microbiota, higher IgA concentrations in the feces and serum, and increased IgA$^+$ plasma cell frequencies in the cLP[309]. IgA controls the composition of gut microbiota by promoting colonization with mutualistic species and neutralizing invasive pathogens[436]. This increase in IgA responses in absence of c-Maf is likely due to the impaired differentiation of Tfr cells in Peyer's patches[305,309], which are critical for IgA selection. Additionally, IL-10$^{Treg\ KO}$ mice showed increased fecal IgA concentrations[309], suggesting that impaired IL-10 production in absence of c-Maf also contributes to the dysregulation of IgA secretion.

Consequent to the dysregulation of multiple mechanisms responsible for the control of microbiota composition, c-Maf deletion in Treg cells leads to the development of dysbiosis in one study[309]. Interestingly, colonizing WT germ-free mice with the microbiota from c-Maf$^{Treg\ KO}$ mice was sufficient to drive excessive intestinal Th17 responses, despite exhibiting normal RORγt$^+$ Treg and Tfr populations[309]. This suggests the existence of a feedback loop, in which Treg cells upregulate c-Maf in response to the microbiota, which in turn enables them to shape and maintain intestinal microbiota composition. This phenomenon could explain the gradation in colitis development observed across different studies. Indeed, while we observed the development of early spontaneous colitis, c-Maf$^{Treg\ KO}$ mice did not develop colitis in two facilities[305,309], and developed late-onset colitis in two other facilities[307,308]. The disparity in the intestinal phenotype of c-Maf$^{Treg\ KO}$ mice across different studies could relate to differences in the microbiota composition of mice bred in distinct animal facilities, leading to different levels of basal inflammation. It would therefore be interesting to compare the microbiota composition of c-Maf$^{Treg\ KO}$ mice across all facilities to substantiate this claim.

Thus, although the exact role of c-Maf in the suppressive capacity of intestinal Treg cells in distinct pathophysiological settings awaits further investigation, it is clear that this transcription factor is a key player in the control of intestinal homeostasis and in the tolerance to gut pathobionts, thereby protecting the host from the development of chronic intestinal inflammation.

4.2. Multiple signaling pathways control intestinal Treg subsets

The existence of distinct tissular Treg specialization programs suggests that environmental signals are involved in the specialization of Treg function. As evidenced by the high expression of c-Maf among eTreg cells, cell activation itself is a driver of c-Maf expression in Treg cells. Indeed, *in vitro* activation and co-stimulation of splenic naïve Treg cells was sufficient to induce c-Maf expression in a proportion of cells[309]. c-Maf expression is enriched in intestinal Treg cells, in effector sites (siLP, cLP), as well as induction sites (mLN, PP). Wheaton et al. hypothesized that the enrichment of c-Maf expression in intestinal Treg cells reflects the abundance of activated Treg cells in the intestine[305]. Regardless, intestinal eTreg cells exhibit higher c-Maf expression levels compared to splenic eTreg cells[309], suggesting that additional signals present in the gut microenvironment contributed to c-Maf expression.

The expression of RORγt in Treg cells being tightly correlated to that of c-Maf, we investigated the signaling pathways involved in the expression of both transcription factors using both *in vivo* and *in vitro* approaches. Strikingly, while the expression of RORγt was decreased in most *in vivo* models used in this study, we could not identify situations where c-Maf expression was lost or considerably decreased in Treg cells *in vivo*. Microbial signals, IL-6/STAT3, and TGF-β signaling cooperate to induce optimal RORγt expression in Treg cells *in vivo*. Disruption of any one of these signals considerably decreases RORγt expression in Treg cells. While c-Maf expression in Treg cells is only partially reduced in absence of STAT3 *in vivo*, the induction of c-Maf relies mainly on IL-6/STAT3 signaling *in vitro*. We also showed that TGF-β and SCFA are able to induce c-Maf expression in Treg cells *in vitro*. However, the disruption of these signals did not decrease c-Maf expression *in vivo*. Contrary to Neumann et al., who reported an important decrease of c-Maf expression in germ-free mice[309], we only observed a moderate decrease of c-Maf expression in germ-free mice, mainly due to the loss of RORγt⁺ Treg cells. Neumann et al. also showed that activating the Notch pathway in splenic Treg cells increased c-Maf expression and that Notch-deficient mice exhibited decreased c-Maf expression in colonic

Treg cells[309]. These results indicate that, while c-Maf integrates many signaling pathways in seemingly redundant ways, RORγt induction is more tightly restricted and requires a combination of intestinal signals. This also shows that c-Maf is necessary but not sufficient for the indunction of RORγt in Treg cells.

The observation that overexpression of c-Maf in *in vitro*-polarized Treg cells is sufficient to induce RORγt expression contrasts with the abundance of c-Maf⁺ RORγt⁻ Treg cells found *in vivo*. This suggests a negative regulation of RORγt expression operating downstream or independently of c-Maf *in vivo*. We consequently found that STAT1 signaling negatively regulates RORγt expression independently of c-Maf. This aspect is illustrated by the dual role of IL-27, which acts as an inhibitor of RORγt, in a STAT1-depependent manner, and as an inducer of c-Maf, in a STAT3-dependent manner. In the context of Th1 intestinal inflammation, during *Toxoplasma gondii* infection, intestinal Treg cells lose expression of RORγt, in a STAT1-dependent manner. This downregulation occurs in presence of an active IL-6/STAT3 pathway, suggesting a dominant negative effect of STAT1 over STAT3 in this process. We will further study the interplay between STAT3 and STAT1 in the expression of RORγt in Treg cells. In particular, we will test the ability of either transcription factor to bind, transactivate, or inactivate, the *Rorc* locus in Treg cells, individually or as part of STAT3/STAT1 heterodimers. Interestingly, while intestinal Treg cells transition from a RORγt⁺ phenotype to a T-bet⁺ phenotype, Treg cells maintain c-Maf expression during Th1 inflammation. It would be interesting to evaluate the impact of a post-infectious phenotypical Treg switch over microbial tolerance and the development of post-infectious intestinal inflammation, as well as the role played by c-Maf in this process. A summary of our data regarding the molecular pathways involved in the induction of c-Maf and RORγt expression in Treg cells is presented in **Figure 43**. Understanding how environmental signals are integrated and how they regulate the specialization of intestinal Treg cells could allow us to specifically impact intestinal tolerance and lead to therapeutic advancements in the context of inflammatory bowel disease.

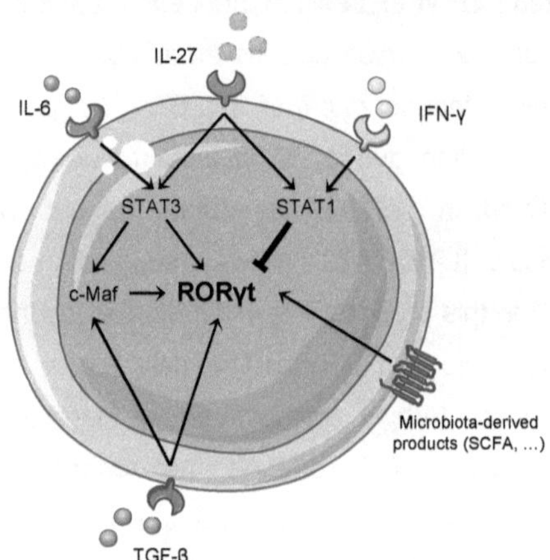

Figure 43. **c-Maf and RORγt induction in Treg cells.** The differentiation of specialized cell subsets is a balancing act: multiple positive and negative signals are integrated in order to tailor cell specialization to the immune context. Signals derived from a complex microbiota, or IL-6/STAT3 and TGF-β signaling induce RORγt expression in Treg cells in a c-Maf-dependent or independent fashion. Contrastingly, inflammatory STAT1 signaling opposes RORγt expression in Treg cells in a c-Maf independent fashion.

4.3. c-Maf⁺ Treg cells promote colon cancer development during inflammation

A recent retrospective study on muscle-invasive bladder cancer patients revealed that highly suppressive tumor infiltrating CCR8⁺ Treg cells exhibit increased levels of c-Maf expression[437]. In line with the high levels of c-Maf in eTreg cells, c-Maf expression is enriched in tumor-infiltrating Treg cells in ectopic models of thymoma (EG7) and colon carcinoma (MC38). The deletion of c-Maf in Treg cells does not however impact tumor development in these models, indicating that c-Maf is not required for the control of tumor development by Treg cells in ectopic models. Having previously shown the role of c-Maf⁺ Treg cells in the control of intestinal immune responses, we investigated the role of c-Maf in the development of colitis-induced colon cancer (CAC). In the AOM/DSS model, which recapitulates human CAC[438], polyps develop as a result of inflammation-induced proliferation of transformed cells. Interestingly, and contrary to our initial predictions, c-Maf[Treg KO] mice treated with AOM/DSS develop fewer polyps, despite showing exacerbated Th17 intestinal inflammation, with increased IL-17A and decreased IL-10 production. These results highlight the dual role of c-Maf-expressing Treg cells in intestinal immunity. c-Maf is required for the control of homeostatic Th17 responses by Treg cells and contributes to the maintenance of intestinal homeostasis. Oppositely, in the context of intestinal tumor development, c-Maf expression in Treg cells favors tumor development. This part of our project is still in progress and will be pursued further in the course of an upcoming PhD book. **Figure 44** summarizes our current hypotheses.

While this was not our case, Neumann et al. observed that the increased IL-17 and IL-22 expression found in c-Maf[Treg KO] mice protected them from acute DSS inflammation[309]. It would therefore be interesting to better characterize the immune response of c-Maf[Treg KO] mice after acute/chronic DSS treatment and after AOM/DSS treatment, to identify the subsets that could underlie AOM/DSS protection. The ability of immune cells to infiltrate the tumor is essential for effective tumor rejection[439], which would support the analysis of the immune infiltrate within tumors. Previous RNA-seq analyses on c-Maf-deficient Treg cells have shown that c-Maf deficiency alters the expression of molecules involved in cell localization[305,307,309], which might affect the ability of Treg cells to migrate and infiltrate the

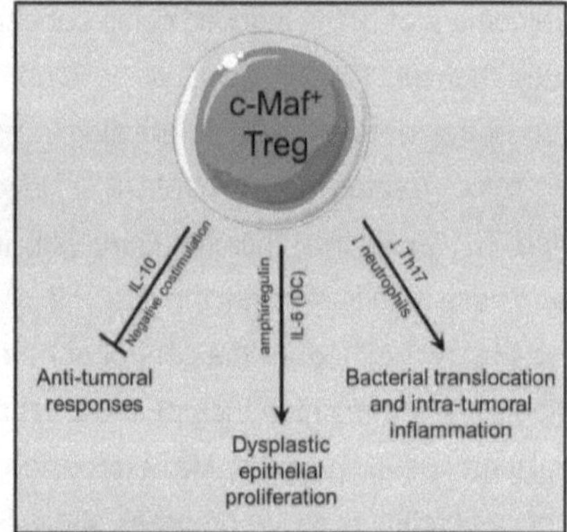

Figure 44. **c-Maf plays dual role in gut homeostasis.** While c-Maf expression in Treg cells is beneficial for the maintenance of gut homeostasis at steady state, it promotes the development of polyps in a colitis-associated colon cancer model. The mechanisms underlying this process remain to be identified and are summarized above.

tumor. We could for instance assess the location of Treg cells with regards to the tumor and to other immune cell types involved in the anti-tumor response by immunofluorescence.

Interestingly, Th17 inflammation, particularly, IL-17A and IL-22 production, has been shown to contribute to the formation, growth, and metastasis of a wide range of tumors, including colon cancer[440–443]. This is contradictory with the phenotype observed in c-Maf[Treg KO] mice, which show increased Th17 inflammation concomitant with a lower tumor load. The role of Th17 inflammation in the phenotype of c-Maf[Treg KO] mice could be assessed by treating mice with anti-IL-17A mAbs. While Th17 inflammation is habitually portrayed as pro-tumoral, Th17-associated cytokines could also hinder tumor development via their effect on myeloid cells[444,445]. In particular, Dmitrieva-Posocco et al reported that neutrophils prevent the infiltration of bacteria within the tumor and thereby dampen tumor-promoting inflammation[444]. We propose to examine the infiltration of neutrophils within the tumors of c-Maf[Treg KO] mice and to study their role by treating mice with anti-Ly6c mAbs.

Dysbiosis is known to promote inflammation and tumor development in mice[446]. As previously stated, the microbiota of c-Maf[Treg KO] mice is sufficient to induce exacerbated Th17 inflammation[309]. We thus propose to assess the role of the microbiota in the phenotype of c-Maf[Treg KO] mice by treating mice with antibiotics or colonizing germ-free mice with fecal bacteria from WT or c-Maf[Treg KO] mice prior to AOM/DSS tumor induction.

c-Maf-deficient Treg cells show a decreased production of IL-10, which is an essential mechanism of mucosal immune suppression. An altered IL-10 Treg production could in turn enhance anti-tumoral immunity, thus hindering tumor development in c-Maf[Treg KO] mice. However, the distinct phenotypes of AOM/DSS-treated IL-10 and c-Maf[Treg KO] mice suggest that c-Maf⁺ Treg cells could sustain tumor development partly through IL-10-independent pathways. Indeed, contrary to c-Maf[Treg KO] mice, IL-10[Treg KO] mice showed enhanced tumor development, possibly driven by a colonic over-inflammation[447]. Indeed, RORγt⁺ Treg cells have recently been shown to sustain IL-6 production by dendritic cells, thereby triggering a STAT3-dependent proliferation of neoplastic epithelial cells and promoting polyp formation[448]. This deleterious effect of RORγt⁺ Treg cells could be lost in absence of c-Maf, as these cells cannot differentiate, and could therefore account for part of the observed phenotype.

Of note, tissue-infiltrating activated Treg cells have been shown to promote malignancy through the direct stimulation of epithelial cell growth via amphiregulin[329].This potential pro-tumoral role of c-Maf-expressing Treg cells could be investigated as transcriptional analysis showed that c-Maf$^+$ Treg cells show high *Areg* expression[307]. A single cell RNA sequencing analysis of Treg cells sorted from the colon of tumor-bearing and tumor-free mice, both of wild type and c-Maf$^{Treg\ KO}$ genotypes, could further allow us to identify immunoregulatory genes as well as non-immune genes targeted by c-Maf in Treg cells.

This work will help identify the immune mechanisms involved in the control of colorectal tumor progression and could pave the way towards novel immune targeted treatments.

Chapter 5: Conclusion

The aim of this work was to determine the role of transcription factor c-Maf in the differentiation and function of Treg cells, particularly in the control of Treg functional and tissular specialization. The results of this work are summarized graphically in **Figure 45**.

We demonstrated the essential role of transcription factor c-Maf in the control of gut homeostasis. c-Maf is required for Treg cells to control homeostatic Th17 responses in the intestine. While c-Maf and RORγt expression are tightly linked in intestinal Treg cells, c-Maf expression is necessary but not sufficient for the induction of RORγt in Treg cells, which relies on multiple signaling pathways. We also reported that RORγt expression is negatively regulated by STAT1 signaling, revealing a potential role of Th1 inflammation in constraining RORγt expression in Treg cells.

Beyond its role in the control of RORγt expression, we showed a previously unknown role for c-Maf in the control of IL-10 expression in intestinal Treg cells. As c-Maf notably controls IL-10 expression in many abundant intestinal immune subsets, this establishes c-Maf as a key regulator of intestinal IL-10 production and further confirms the role of c-Maf in the control of gut homeostasis.

Finally, we showed that, despite an exacerbated Th17 inflammation, c-Maf deficiency in Treg cells inhibits the development of colitis-associated colon cancer, indicating that c-Maf exerts a specific role in the control of colon tumor immunity.

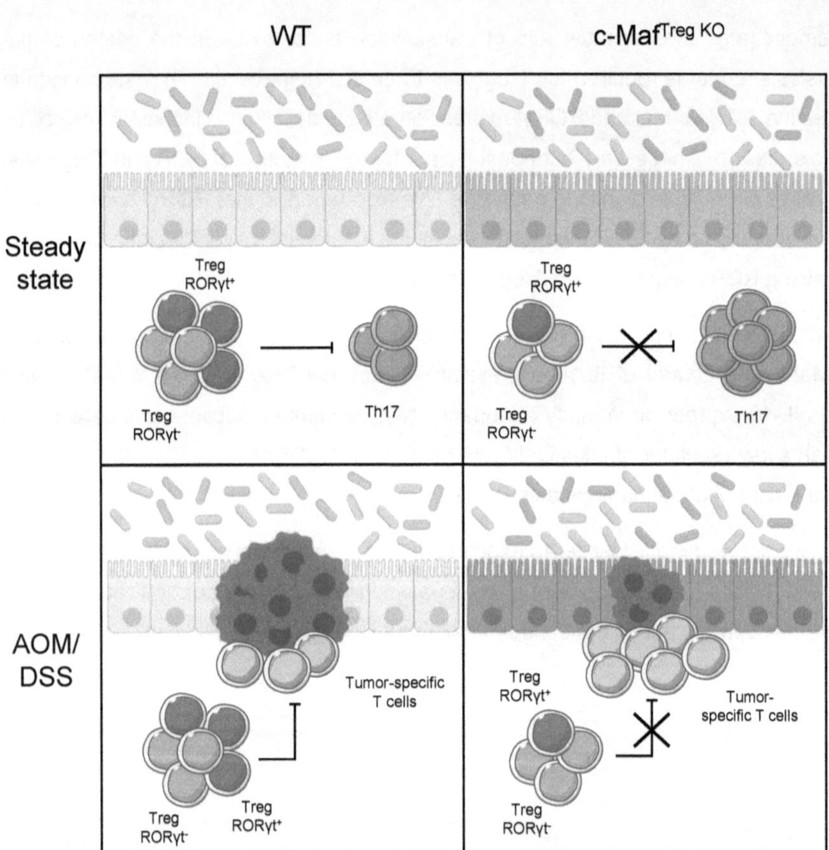

Figure 45. **Graphical abstract.** At steady state, c-Maf+ Treg cells control Th17 intestinal response and thus promote tolerance to the microbiota and gut homeostasis. In the context of colitis-induced colon cancer, c-Maf expression in Treg cells is deleterious for the host as it promotes tumor development.

Chapter 6: Materials and methods

6.1. Mice

The mouse strains used during this work are summarized in **Table 1**. The housing conditions of the respective mouse strains are summarized in **Table 2**. c-Maf-flox mice were crossed with CD4^{Cre} mice or Foxp3^{CreYFP} mice, to generate mice deficient for c-Maf specifically in T cells[‡] (c-Maf$^{CD4\ KO}$) or Foxp3$^+$ Treg cells (c-Maf$^{Treg\ KO}$), respectively. STAT3-flox mice were crossed with CD4^{Cre} mice to generate mice deficient for STAT3 specifically in T cells (STAT3$^{CD4\ KO}$). Tgfbr2-flox mice on a NOD background, crossed with Foxp3^{Cre} mice (TGFbRII$^{Treg\ KO}$), were housed and bred at the UCSF Animal Barrier Facility.

All mice were used between 6 and 12 weeks of age. The experiments were carried out in compliance with the relevant laws and institutional guidelines and were approved by the Université Libre de Bruxelles Institutional Animal Care and Use Committee (protocol number CEBEA-4).

6.2. Solutions and media

Reagent and media references are summarized in **Table 3**.

6.2.1. General solutions and media

- **Complete RPMI medium:** Roswell Park Memorial Institute Medium 1640, 5.10^5 M β-mercaptoethanol, 4% (1mM sodium pyruvate, 2mM L-glutamine, 5000U/mL penicillin/5000µg/mL streptomycin, 1% non-essential amino acids). Complete medium is supplemented with different fetal calf serum (FCS) content. The medium most often used for cell culture is RPMI 5% FCS.
- **PBS buffer:** Phosphate Buffered Saline (1X), without Calcium and Magnesium.
- **FACS medium:** PBS, 0,1% BSA, 0,1% azide.
- **MACS medium:** PBS, 0,5% BSA, 2mM EDTA.
- **Red blood cell lysis buffer (ACK):** 90% NH$_4$Cl 0,16M, 10% Tris buffer 0,17M, pH 7,65.

[‡] CD4 is expressed in every T cell during the differentiation process, at the double positive stage. CD4^{Cre} mice will therefore delete the targeted gene in every T cell, including CD8 T cells.

Strain	MGI ID	Genetic background	Origin
WT		C57BL/6	Envigo (Horst, The Netherlands)
c-Maf flox	5316775	C57BL/6	C. Birchmeier (Max Delbrück Center for Molecular Medicine, Berlin, Germany)
Foxp3 YFP-Cre	3790499	C57BL/6	A. Liston (KU Leuven, Leuven, Belgium)
CD4-CRE	2386448	C57BL/6	G. Van Loo (Ghent University, Ghent, Belgium)
IL-6 KO	1857197	C57BL/6	Jackson Laboratory (Bar Harbor, ME, USA)
STAT3 flox/flox	1926816	C57BL/6	S. Akira (Osaka University, Osaka, Japan)
FoxP3-GFP-hCre	3809724	NOD	Q. Tang (University of California San Francisco, SF, USA)
Tgfbr2 flox FoxP3-GFP-hCre	2384513, 3809724		
Germ free		C57BL/6	Ghent Germfree and Gnotobiotic mouse facility (Ghent University, Ghent, Belgium)
STAT1 KO	1930947	C57BL/6	D.E. Levy (New York University School of Medicine, NYC, USA)

Table 1. **Mouse strains and provenance.**

Strain	Housing conditions
WT (c-Maf flox)	IVC
c-Maf CD4 KO	
WT (Foxp3 YFP-Cre)	IVC
c-Maf Treg KO	
WT	SPF
IL-6 KO	
WT (STAT3 flox)	SPF
STAT3 CD4 KO	
WT (FoxP3-GFP-hCre)	SPF*
Tgfbr2 Treg KO	
WT	SPF
Germ free	Germ free facility**
WT	SPF
STAT1 KO	

Table 2. **Housing conditions of mouse strains.** *UCSF Animal Barrier Facility (University of California San Francisco, SF, USA). **Ghent Germ free and Gnotobiotic mouse facility (Ghent University, Ghent, Belgium).

Reagent	Abbreviation	Company	Reference
Ampicillin		Sigma-Aldrich	A9518
Anti-CD3 antibody		BioXcell	145-2c11
Anti-CD4 MACS beads		Miltenyi	130-117-043
Anti-CD90.2 MACS beads		Miltenyi	130-121-278
Anti-ICOSL antibody (HK5.3)	α-ICOSL	BioXcell	BE0028
Azide		Merck	822335
Azoxymethane	AOM	Sigma-Aldrich	A5486
Bovine Serum Albumin	BSA	Sigma-Aldrich	
Brefeldin A	BfA	eBioscience	00-4506-51
Carboxyfluorescein	CFSE	ThermoFisher	C34554
Collagenase, type 3		Worthington	LS004180
Dextran Sulfate Sodium salt, colitis grade (36,000 - 50,000 Da)	DSS	MP Biomedical	160110
Dithiothreitol	DTT	Sigma-Aldrich	D0632
DNase I		Roche	10104159001
Fetal Calf Serum	FCS	Sigma-Aldrich	F0804
Formaldehyde	FA	VWR Chemicals	10790-708
FOXP3 staining buffer set		eBioscience	00-5523-00
Hank's Buffered Saline Solution without Calcium and Magnesium, with Phenol Red	HBSS	Lonza	10-543Q
Hepes		Gibco	15630080
Ionomycin	Iono	Sigma-Aldrich	I0634
L-glutamine	L-Glut	Gibco	25030081
Liberase		Roche	05401020001
Lymphoprep		StemCELL	07851
Methanol	MetOH	VWR Chemicals	20847-320
Metronidazole		Duchefa	M0131
Neomycin		Sigma-Aldrich	N1876
Non-essential amino acids	NEAA	Gibco	11140050
NucleoSpin RNA Plus Kit		Macherey-Nagel	15370195
Penicillin/streptomycin	Pen/strep	Gibco	15070063
Pentobarbital sodium		Ceva	165A2
Percoll		GE Healthcare	17-0891-02
Phorbol 12-myristate 13-acetate	PMA	Sigma-Aldrich, P8139	P8139
Phosphate Buffered Saline (1X), without Calcium and Magnesium	PBS	Lonza	17-516Q
Phosphate Buffered Saline (10X), without Calcium and Magnesium	PBS	Lonza	17-515F
Polybrene		Gibco	TR-1003
RNeasy Mini Kit		Qiagen	74106
Roswell Park Memorial Institute Medium 1640	RPMI	Lonza	BE12-167F
Sodium pyruvate	NaPyr	Gibco	11360070
Superscript II reverse transcriptase		Invitrogen	18064022
SYBR Green Master mix kit	SYBR Green	ThermoFisher	4309155
TRIzol		Invitrogen	15596026
Vancomycin		Duchefa	V0155
β-mercaptoethanol	β-MEtOH	Sigma-Aldrich	M6250

Table 3. **Reagent references.**

- **PMA/ionomycin/Brefeldin A:** complete RPMI medium, 10% FCS, 50ng/mL Phorbol 12-myristate 13-acetate (PMA), 250ng/mL ionomycin, and x1000 Brefeldin A.
- **Collagenase/DNase solution:** HBSS buffer without Phenol Red, 200U collagenase, 40µg/mL DNase I.
- **40%/70% Percoll solution:** 40 or 70% of (9/10 Percoll, 1/10 PBS 10X), complete with PBS 1X.

6.2.2. Isolation of intestinal leukocytes

- **RPMI medium (gut):** complete RPMI medium, 20mM Hepes.
- **HBSS buffer (gut):** Hank's Buffered Saline Solution without Calcium and Magnesium, with Phenol Red, 20mM Hepes.
- **HBSS 3% FCS:** HBSS buffer (gut) complemented with 3% FCS.
- **HBSS 3% FCS/DTT/EDTA:** HBSS buffer (gut) complemented with 3% FCS, 72,5µg/mL DTT, 2,5mM EDTA.
- **HBSS/EDTA:** HBSS buffer (gut), 2,5mM EDTA.
- **RPMI/liberase/DNase:** RPMI medium (gut), 400µg/mL DNase I, 20µg/mL liberase.
- **RPMI 3% FCS/EDTA:** RPMI medium (gut), 3% FCS, 2,5mM EDTA.
- **30% Percoll solution:** 30% Percoll, 70% RPMI medium (gut).

6.3. Antibodies, intracellular staining, and flow cytometry

Antibody references and concentrations are summarized in **Table 4**. To exclude dead cells, Live/Dead fixable near-IR stain was used for Canto II analysis and iFluor 860 maleimide was used for CytoFLEX analysis. These stains were not used for phosphorylation staining.

For cytokine staining, cells were resuspended in 100µL cell culture medium (RPMI 10% FCS) and stimulated with 100µL PMA/ionomycin/Brefeldin A solution in a round-bottomed 96 well plate for three hours.

All cell incubations with fluorescently-conjugated antibodies were done in the dark. For transcription factor and cytokine staining, cells were stained in round-bottomed 96 well plates. After being washed twice with PBS, cells were incubated 25min with 50µL extracellular staining, diluted in PBS, at 4°C. Cells were washed twice in PBS and fixed for 30min in 100µL Fixation/Permeabilization solution (¾ diluent, ¼ concentrate; FOXP3 staining buffer set) at 4°C. Cells were washed twice with Permeabilization buffer and incubated for 25min with 50µL intracellular staining, diluted in Permeabilization buffer, at

	Target/ reagent	Clone	Fluorochrome (or equivalent)	Company	Reference	Concentration	
Extracellular	CD25	7D4	PE	Miltenyi	130-118-550	1/100	
	CD278	C398.4A (ICOS)	Biotin	eBioscience	13-9949-82	1/100	
	CD304	3DS304M (Nrp-1)	PerCP eF710	eBioscience	46-3041-80	1/100	
	CD4	RM4-5	A700	BD Biosciences	557956	3/500	
	CD4	RM4-5	Pacific Blue	BD Biosciences	558107	1/100	
	CD44	IM7	PECy7	BD Biosciences	560569	1/100	
	CD62L	MEL-14	A700	BD Biosciences	560517	1/100	
	Live/Dead		APC-Cy7	Invitrogen	L10119	1/100	
	Streptavidin		PECy7	BD Biosciences	557598	1/100	
	TCRb	H57-597	PerCP-Cy5.5	BD Biosciences	560657	1/100	
	α-FcR (blocking)			BioXcell	BE0307	1/400	
	Ly6G	1A8	BUV395	BD Biosciences	565964	1/100	General immune cell panel
	CD4	GK1.5	BUV496	BD Biosciences	564667	1/100	
	CD44	IM7	BUV737	BD Biosciences	564392	1/50	
	CD161	PK136	BV421	Biolegend	108731	1/50	
	CD11c	HL3	BV480	BD Biosciences	565627	1/50	
	CD62L	MEL-14	BV650	BD Biosciences	564108	2/50	
	CD11b	M1/70	BV711	BD Biosciences	563168	1/200	
	CD25	PC61	BB515	BD Biosciences	564424	1/100	
	SiglecF	E50-2440	PECF594	BD Biosciences	562757	1/100	
	CD8a	53-6.7	PECF594	BD Biosciences	562315	1/100	
	CD45R	RA3-6B2	PerCP-C5.5	BD Biosciences	561101	2/50	
	CD49d	R1-2 (DX5)	PE	eBioscience	12-0492-82	2/50	
	TCRgd	eBioGL3 (GL-3, GL3)	PECy5	eBioscience	15-5711-82	1/50	
	CD3e	145-2C11	Biotin	eBioscience	13-0031-82	1/100	
	CD19	1D3	PECy7	BD Biosciences	561739	1/200	
	CD317	BST2, PDCA-1	APC	Biolegend	127016	2/50	
	F4/80	T45-2342	APCR700	BD Biosciences	565787	1/100	
	Ly6c	AL-21	APCCy7	BD Biosciences	560596	1/100	
	MHCII	34-5-3	CF790	BD Biosciences	553611	1/50	
	Maleimide		iFluor860	AAT Bioquest	1408	1/1000	
	Streptavidin		PECy5	eBioscience	15-4317-82	1/100	
pSTAT	STAT1	pY701	Alexa 488	BD Biosciences	612596	1/10	
	STAT3	4/P-STAT3 (Tyrosine)	Alexa 647	BD Biosciences	557815	1/10	
	STAT5	47/Stat5 (pY694)	PE	BD Biosciences	612567	1/10	
Intracellular	c-Maf	symOF1	EF660	eBioscience	50-9855-82	3/500	
	Foxp3	FJK-16s	FITC	eBioscience	22-5773-82	1/100	
	GATA3	L50-823	PE	BD Biosciences	560074	1/10	
	RORγt	B2D	PE	eBioscience	12-6981-82	1/100	
	RORγt	Q31-378	PECF594	BD Biosciences	562684	1/200	
	T-bet	eBio4B10 (4B10, 4-B10)	PE	eBioscience	12-5825-82	1/100	
Cytokines	IL-10	JES5-16E3	APC	BD Biosciences	554468	1/100	
	IL-10	JES5-16E3	BV421	BD Biosciences	566295	1/100	

Table 4. **Antibodies and reagents used for flow cytometry.**

4°C. Cells were washed twice in Permeabilization buffer and resuspended in PBS/FACS buffer.

For phosphorylation staining, cell suspensions were placed in Eppendorf tubes, spun down, and fixed 10min in a 37°C water bath with 1mL 1% formaldehyde. Cells were washed twice with PBS and permeabilized 30min on ice with 1mL cold methanol. Cells were washed twice with FACS medium and incubated 25min with the staining solution containing antibodies for surface markers, phosphorylated residues and intracellular markers.

Flow cytometric analysis was performed on a Canto II (BD Biosciences) or CytoFLEX (Beckman Coulter) and analyzed using FlowJo software (Tree Star).

6.4. Leukocyte purification

6.4.1. Intestinal tissues

After removal of Peyer's patches and mesenteric fat, intestinal tissues were cut longitudinally and manually washed with cold HBSS 3% FCS followed by cold PBS. Intestinal tissues were cut in small sections and incubated in 50mL Falcon tubes containing 20mL HBSS 3% FCS/EDTA/DTT for 30min at 37°C with agitation (200rpm). After incubation, intestinal tissues were strained and washed three times with 20mL HBSS/EDTA and once with HBSS to remove epithelial cells. Intestinal tissues were then minced in beakers containing 5mL RPMI/liberase/DNase, topped up with 10mL of the same solution and incubated at 37°C for 30min (small intestine) or 45min (colon). Intestinal tissues were neutralized with 20mL RPMI 3%/EDTA and then mashed on a 70µm cell strainer placed on 50mL Falcon tubes and rinsed with RPMI 3%/EDTA to reach 50mL. After centrifugation, cells were resuspended in 10mL 30% Percoll solution, placed in a 15mL Falcon tube and centrifuged for 15min at 1600rpm at 20°C, without brakes. Cells were then resuspended at the appropriate concentration.

6.4.2. Secondary lymphoid organs

Lymph nodes and thymus were mechanically disrupted in cell culture medium. Spleens were mechanically disrupted in red blood cell lysis buffer (1mL per spleen), which was neutralized with RPMI 5% FCS. Lymph nodes, thymus, and spleen were strained, centrifuged 7min at 1400 rpm at 4°C and resuspended at the appropriate concentration.

6.4.3. Liver and lung

Mice were anesthetized with 100µL/20g Pentobarbital sodium, injected intraperitoneally. After anesthesia, mice were perfused with PBS. Liver and lung samples were injected with 3mL of a collagenase/DNase solution, incubated at 37°C for 10min and mechanically disrupted. Lung and liver samples were neutralized with RPMI 5%/EDTA, strained and washed. After centrifugation, cells were resuspended in 6mL of a 40% Percoll solution and placed on 2mL 70% Percoll solution in a 15mL Falcon tube. Cells were centrifuged for 15min at 2000rpm at 20°C, without brakes. Leukocytes were collected at the interphase, washed in cell culture medium and resuspended at the appropriate concentration.

6.5. Acute DSS-induced colitis

Mice were administered with 2% Dextran Sulfate Sodium salt (colitis grade: 36,000 - 50,000 Da) diluted in drinking water for 7 days. Weight loss and survival rates were assessed every other day for 2 weeks.

6.6. Anti-CD3-induced enteritis

Mice were injected with 15µg anti-CD3 purified antibody intraperitoneally. Weight loss and survival rates were assessed every other day for 10 days.

6.7. *Toxoplasma gondii* infection

ME-49 type II *T. gondii* was kindly provided by Dr. De Craeye (Institut Scientifique de Santé Publique, Belgium). C57BL/6 mice were inoculated with three cysts by gavage, to generate chronically infected mice, leading to the production of tissue cysts 1-3 months later. Tissue cysts were extracted from brain tissue and counted. Mice were subsequently infected with 10 cysts in 200µL PBS by intragastric gavage and sacrificed at day 8 after infection for FACS analysis, or monitored for survival until day 21.

6.8. Antibiotics treatment

Mice were treated with wide spectrum antibiotics for 3-4 weeks. Control mice were administered with sweetened drinking water, which was supplemented with 1g/L ampicillin, 1g/L neomycin, 0.5g/L vancomycin and 1g/L metronidazole for the treated group. The size of the cecum was compared between control and treated groups to validate the efficacy of the treatment.

6.9. T cell culture

Organs were isolated as described in **6.4.2.** Splenocytes (unless otherwise mentioned) were resuspended in incubated at 4°C with anti-CD4 MACS beads (40µL/spleen) and 90µL MACS buffer. After 10min, the antibody mix was added (100µL/spleen) and cells were incubated for an additional 10min. Cells were washed and CD4$^+$ T cells were positively selected by passing the cell suspension through a pre-conditioned MACS LS column (Miltenyi) placed in a QuadroMACS™ Separator (Miltenyi) according to the product protocol. Cells were washed in cell culture medium and strained. Naïve CD4 T cells were sorted by flow cytometry as CD4$^+$ YFP$^-$/CD25$^-$ CD62L$^+$ CD44$^-$ cells (**Table 5**). After sorting, cells were washed in cell culture medium and resuspended at the appropriate concentration (usually 5.10^5 cells in 24 well plates and $7,5.10^4$ cells in 96 well plates).

Cultures plates were coated with 5µg/mL anti-CD3 diluted in PBS at 37°C for 1.5 hours, then washed with cell culture medium. Cells were cultured 72 hours in 24 or 96 well plates. The reagents used for cell culture are summarized in **Table 6**. To generate iTreg cells, TGF-β was used alone, in combination with IL-2 (Treg condition) or in combination with both IL-2 and IL-6 (Tr17 condition).

6.10. Retroviral infection

Platinum-E retroviral packaging cells (T. Kitamura, University of Tokyo, Tokyo, Japan) were transfected with a c-Maf encoding retroviral plasmid (pMIEG-c-Maf-IRES-GFP) or a

Target	Clone	Fluorochrome	Concentration
Anti-FcR			1/400
CD4	RM4-5	Pacific Blue	1/100
CD44	IM7	PECy7	1/100
CD62L	MEL-14	A700	1/100
CD25	PC61	BB515	1/100

Table 5. **Antibody cocktail for MACS purification of naïve CD4 T cells.** For YFP-Cre mice, CD25 was replaced by YFP expression as a Treg marker.

Reagent	Concentration	Company	Reference
Anti-CD3	5µg/mL (coated)	BioXcell	BE0001-1
Anti-CD28	1µg/mL	BioXcell	BE0328
TGF-β	2-5ng/mL	eBioscience	14-8342-62
IL-2	10ng/mL	Peprotech	212-12
IL-6	10ng/mL	eBioscience	216-16
Acetate	10mM	Sigma-Aldrich	
Propionate	0.5mM	Sigma-Aldrich	
Butyrate	0.125mM	Sigma-Aldrich	
IFN-γ	10, 100ng/mL	Peprotech	315-05
IL-27	50ng/mL	Biolegend	577402
Anti-IL-4	10µg/mL	BioXcell	BE0045
Anti-IFN-γ	10µg/mL	BioXcell	BE0312

Table 6. **Cytokines and reagents used for CD4 T cell polarization.**

control retroviral plasmid (pMIEG-IRES-GFP) to produce retrovirus-containing supernatants. 24 hours after activation, naïve CD4 T cells, polarized as indicated, were infected during a 90-min centrifugation with 1mL retrovirus-containing supernatant and polybrene. 48 hours later, infected cells were FACS-sorted based on GFP expression and were stimulated with anti-CD3 for 24 hours (5µg/mL, coated) before flow cytometry staining.

6.11. Treg cell *in vitro* suppression assay

Splenocytes from Foxp3 YFP[Cre] and c-Maf[Treg KO] mice were enriched in CD4 T cells by positive MACS selection as described in **6.10.** Naïve CD4 T cells from wild-type mice were sorted by flow cytometry as CD4$^+$ CD44^{lo} CD62L^{hi} YFP$^-$ cells and subsequently labeled with CFSE. WT or c-Maf-deficient Treg cells from spleen, mLN, or *in vitro* polarized Tr17 cells were sorted by flow cytometry as CD4$^+$ YFP$^+$ cells. Splenocytes from wild-type C57BL/6 mice were depleted in T cells (anti-CD90.2 beads) using MACS LS columns (Miltenyi) to collect APCs. 4×10^4 CFSE-labeled naïve CD4 T cells were cultured for 72h with APCs (1×10^5) and soluble anti-CD3 (0.5µg/mL) in the presence or absence of Treg cells, as indicated. Cell proliferation of naïve CD4 T cells was analyzed by FACS.

6.12. RT-qPCR

mLN RNA was extracted using the TRIzol method. siLP and cLP RNA was extracted with NucleoSpin Plus kit. Intestinal tissues, placed in 1mL lysis buffer in 14mL polypropylene tubes, were first homogenized with polytron at full speed. RNA from intestinal tissues was then extracted following the manufacturer's instructions (using 350µL of homogenate). RNA was reverse transcribed with Superscript II reverse transcriptase according to the manufacturer's instructions. Quantitative real-time RT-PCR was performed using the SYBR Green Master mix kit. Primers were purchased from Sigma-Aldrich. Primer sequences can be found in **Table 7**. mRNA gene expression was relativized with *Rpl32* expression.

6.13. ChIP-seq data

A publicly available ChIP-seq dataset (GSE40918) for c-Maf and Stat3 was downloaded

Target gene	Forward primer	Reverse primer
Ifng	TGCCAAGTTTGAGGTCAACA	GAATCAGCAGCGACTCCTTT
Il10	CCTGGGTGAGAAGCTGAAGA	GCTCCACTGCCTTGCTCTTA
Il12a	CCTCAGTTTGGCCAGGGTC	CAGGTTTCGGGACTGGCTAAG
Il12b	GGAAGCACGGCAGCAGAATA	AACTTGAGGGAGAAGTAGGAATGG
Il17a	ATCCCTCAAAGCTCAGCGTGTC	GGGTCTTCATTGCGGTGGAGAG
Il1b	CAAGCTTCCTTGTGCAAGTG	AGGTGGCATTTCACAGTTGA
Il22	CAGCAGCCATACATCGTCAA	GCCGGACATCTGTGTTGTTA
Il33	TGAGACTCCGTTCTGGCCTC	CTCTTCATGCTTGGTACCCGAT
Il4	ATGCACGGAGATGGATGTG	AATATGCGAAGCACCTTGGA
Il6	GTTCTCTGGGAAATCGTGGA	GCAAGTGCATCATCGTTGTT
Maf	AGCAGTTGGTGACCATGTCG	TGGAGATCTCCTGCTTGAGG
Rorc	TCTACACGGCCCTGGTTCT	ATGTTCCACTCTCCTCTTCTCTTG
Rpl32	ACATCGGTTATGGGAGCAAC	TCCAGCTCCTTGACATTGT
Tgfb	TGACGTCACTGGAGTTGTACGG	GGTTCATGTCATGGATGGTGC
Tnfa	GCCTCCCTCTCATCAGTTCTA	GCTACGACGTGGGCTACAG

Table 7. **qPCR primer sequences.**

and mapped to the mm9 genome using Bowtie2 with sensitive-local predefined parameters. Resulting BAM files were converted to bigwig files and visualized with IGV genome browser.

6.14. Ectopic tumor models

Tumor cells were injected subcutaneously in the flank of the mouse (EG7-OVA: 1.10^6 cells; MC38: 5.10^5 cells). Tumor volume was measured every couple of days. For FACS analysis of tumor tissues, mice were sacrificed between days 15 and 20. Tumor tissue was collected, cut into pieces and digested for 30min in a RPMI/DNase/liberase solution at 37°C. Samples were mechanically disrupted, washed with RPMI 5% FCS/EDTA, and mashed on a 70µm cell strainer. After centrifugation, cells were resuspended in HBSS, strained, placed on top of 3mL room temperature Lymphoprep, and centrifuged at 2000 rpm for 20min at 20°C, without brakes. Leukocytes were collected at the interphase, washed in cell culture medium and resuspended at the appropriate concentration.

6.15. AOM/DSS

Mice were intraperitoneally injected with azoxymethane on day 0. Mice were treated with three courses of 1.2% DSS diluted in drinking water for 5 days, starting from day 3. Between each DSS course, mice were put back on drinking water. Weight and survival were monitored every other day until two weeks after the last DSS course. The experimental procedure is summarized in **Figure 37A**.

6.16. Histopathology

Colons from steady state or AOM/DSS treated Foxp3 YFP^{Cre} and c-Maf$^{Treg\ KO}$ mice were collected, cut longitudinally, and washed with PBS. A 2 cm-long section was put in a tissue cassette and fixed overnight in 4% formalin. Manufacturing of Formalin-Fixed and Paraffin-Embedded (FFPE) blocks as well as slide sectioning and Hematoxylin-Eosin (HE) staining were done at the Center for Microscopy and Molecular Imaging (CMMI). Colon sections were 4µm thick. Histological analysis was conducted by Hanifa

Bouzourene (Unilabs Lausanne, Switzerland). Inflammation and tumor scoring grids are shown in **Tables 8-10**.

6.17. Statistical analysis

For unpaired data, statistical difference between groups was determined by an unpaired t test, when sample size was sufficient and both groups passed the normality test, and by a Mann-Whitney test for two-tailed data otherwise, unless all the values of the control group were identical, in which case a Wilcoxon signed rank test was used. Mutant and control groups did not always have similar standard deviations and an unpaired two-sided Welch's t-test was therefore used. For paired data, a paired t test was used. For data with multiple independent variables, a two-way ANOVA was used. For survival curves, a Mantel-Cox test was used. Error bars represent mean ± SD. No samples were excluded from the analysis.

Description	Score			
Oedema	0	1		
Acute inflammation	0	1	2	3
Chronic inflammation	0	1	2	3
Cryptitis	0	1	2	3
Epithelial erosion	0	1	2	3
Crypt destruction	0	1	2	3
Mucosa	0	1		
Submucosa	0	1		
Muscularis propria	0	1		
Subserosa	0	1		
Lymphangiectasy	0	1		
Lymphoid node	0	1		
Focal/diffuse inflammation	0	1		
Inflammation score				

Table 8. **Intestinal inflammation histological scoring grid.**

Inflammation	Score	Grade
Normal	0	0
Mild	1-5	1
Moderate	5-10	2
Severe	10-16	3

Table 9. **Histological inflammation grade.**

Description	Score
No dysplasia	0
Indefinite low grade due to active inflammation/ulcers	1
Low grade (includes atypical hyperplasia and adenomas)	2
High grade (includes adenomas and non-invasive carcinomas)	3
High grade with invasion beyond tunica muscularis (carcinoma)	4
Tumor score	

Table 10. **Tumor histological scoring grid.**

Chapter 7: References

1. Janeway, C. A. The immune system evolved to discriminate infectious nonself from noninfectious self. *Immunol. Today* **13**, 11–16 (1992).
2. Bosch, T. C. G. & McFall-Ngai, M. J. Metaorganisms as the new frontier. *Zoology* **114**, 185–190 (2011).
3. Belkaid, Y. & Hand, T. W. Role of the Microbiota in Immunity and Inflammation. *Cell* **157**, 121–141 (2014).
4. Moser, M. & Leo, O. Key concepts in immunology. *Vaccine* **28, Supplement 3**, C2–C13 (2010).
5. Dustin, M. L. The immunological synapse. *Cancer Immunol. Res.* **2**, 1023–1033 (2014).
6. Turner, M. D., Nedjai, B., Hurst, T. & Pennington, D. J. Cytokines and chemokines: At the crossroads of cell signalling and inflammatory disease. *Biochim. Biophys. Acta BBA - Mol. Cell Res.* **1843**, 2563–2582 (2014).
7. Altan-Bonnet, G. & Mukherjee, R. Cytokine-mediated communication: a quantitative appraisal of immune complexity. *Nat. Rev. Immunol.* **19**, 205–217 (2019).
8. Eyerich, S., Eyerich, K., Traidl-Hoffmann, C. & Biedermann, T. Cutaneous Barriers and Skin Immunity: Differentiating A Connected Network. *Trends Immunol.* **39**, 315–327 (2018).
9. Parker, D. & Prince, A. Innate immunity in the respiratory epithelium. *Am. J. Respir. Cell Mol. Biol.* **45**, 189–201 (2011).
10. Bustamante-Marin, X. M. & Ostrowski, L. E. Cilia and Mucociliary Clearance. *Cold Spring Harb. Perspect. Biol.* **9**, (2017).
11. Sperandio, B., Fischer, N. & Sansonetti, P. J. Mucosal physical and chemical innate barriers: Lessons from microbial evasion strategies. *Semin. Immunol.* **27**, 111–118 (2015).
12. Okumura, R. & Takeda, K. Maintenance of intestinal homeostasis by mucosal barriers. *Inflamm. Regen.* **38**, 5 (2018).
13. Pickard, J. M., Zeng, M. Y., Caruso, R. & Núñez, G. Gut microbiota: Role in pathogen colonization, immune responses, and inflammatory disease. *Immunol. Rev.* **279**, 70–89 (2017).
14. Bosch, T. C. G. Cnidarian-microbe interactions and the origin of innate immunity in metazoans. *Annu. Rev. Microbiol.* **67**, 499–518 (2013).
15. Muthamilarasan, M. & Prasad, M. Plant innate immunity: an updated insight into defense mechanism. *J. Biosci.* **38**, 433–449 (2013).
16. Riera Romo, M., Pérez-Martínez, D. & Castillo Ferrer, C. Innate immunity in vertebrates: an overview. *Immunology* **148**, 125–139 (2016).
17. Brubaker, S. W., Bonham, K. S., Zanoni, I. & Kagan, J. C. Innate Immune Pattern Recognition: A Cell Biological Perspective. *Annu. Rev. Immunol.* **33**, 257–290 (2015).
18. Takeda, K. & Akira, S. Toll-like receptors. *Curr. Protoc. Immunol.* **109**, 14.12.1-14.12.10 (2015).
19. Shiokawa, M., Yamasaki, S. & Saijo, S. C-type lectin receptors in anti-fungal immunity. *Curr. Opin. Microbiol.* **40**, 123–130 (2017).
20. Kim, Y. K., Shin, J. S. & Nahm, M. H. NOD-Like Receptors in Infection, Immunity, and Diseases. *Yonsei Med. J.* **57**, 5–14 (2016).
21. Kato, H., Oh, S.-W. & Fujita, T. RIG-I-Like Receptors and Type I Interferonopathies. *J. Interferon Cytokine Res. Off. J. Int. Soc. Interferon Cytokine Res.* **37**, 207–213 (2017).
22. Patel, S. Danger-Associated Molecular Patterns (DAMPs): the Derivatives and Triggers of Inflammation. *Curr. Allergy Asthma Rep.* **18**, 63 (2018).
23. Blander, J. M. & Sander, L. E. Beyond pattern recognition: five immune checkpoints for scaling the microbial threat. *Nat. Rev. Immunol.* **12**, 215–225 (2012).
24. Gordon, S. Phagocytosis: An Immunobiologic Process. *Immunity* **44**, 463–475 (2016).
25. Gray, M. & Botelho, R. J. Phagocytosis: Hungry, Hungry Cells. *Methods Mol. Biol. Clifton NJ* **1519**, 1–16 (2017).
26. Lancaster, C. E., Ho, C. Y., Hipolito, V. E. B., Botelho, R. J. & Terebiznik, M. R. Phagocytosis: what's on the menu? *Biochem. Cell Biol. Biochim. Biol. Cell.* **97**, 21–29 (2019).
27. Kolaczkowska, E. & Kubes, P. Neutrophil recruitment and function in health and inflammation. *Nat. Rev. Immunol.* **13**, 159–175 (2013).
28. Nauseef, W. M. & Borregaard, N. Neutrophils at work. *Nat. Immunol.* **15**, 602–611 (2014).
29. Michael, M. & Vermeren, S. A neutrophil-centric view of chemotaxis. *Essays Biochem.* **63**, 607–618 (2019).
30. Liu, X. & Lieberman, J. Knocking 'em Dead: Pore-Forming Proteins in Immune Defense. *Annu. Rev. Immunol.* **38**, 455–485 (2020).
31. Chowdhury, D. & Lieberman, J. Death by a Thousand Cuts: Granzyme Pathways of Programmed Cell Death. *Annu. Rev. Immunol.* **26**, 389–420 (2008).
32. Mancini, M. & Vidal, S. M. Mechanisms of Natural Killer Cell Evasion Through Viral Adaptation. *Annu. Rev. Immunol.* **38**, 511–539 (2020).
33. Holers, V. M. Complement and Its Receptors: New Insights into Human Disease. *Annu. Rev. Immunol.* **32**, 433–459 (2014).
34. West, E. E., Kolev, M. & Kemper, C. Complement and the Regulation of T Cell Responses. *Annu. Rev. Immunol.* **36**, 309–338 (2018).

35. Nielsen, S. C. A. & Boyd, S. D. Human adaptive immune receptor repertoire analysis-Past, present, and future. *Immunol. Rev.* **284**, 9–23 (2018).
36. Attaf, M., Legut, M., Cole, D. K. & Sewell, A. K. The T cell antigen receptor: the Swiss army knife of the immune system. *Clin. Exp. Immunol.* **181**, 1–18 (2015).
37. Clambey, E. T., Davenport, B., Kappler, J. W., Marrack, P. & Homann, D. Molecules in medicine mini review: the αβ T cell receptor. *J. Mol. Med. Berl. Ger.* **92**, 735–741 (2014).
38. Nikolich-Žugich, J., Slifka, M. K. & Messaoudi, I. The many important facets of T-cell repertoire diversity. *Nat. Rev. Immunol.* **4**, 123–132 (2004).
39. Mellman, I. Dendritic cells: master regulators of the immune response. *Cancer Immunol. Res.* **1**, 145–149 (2013).
40. Lindenbergh, M. F. S. & Stoorvogel, W. Antigen Presentation by Extracellular Vesicles from Professional Antigen-Presenting Cells. *Annu. Rev. Immunol.* **36**, 435–459 (2018).
41. Kotsias, F., Cebrian, I. & Alloatti, A. Antigen processing and presentation. *Int. Rev. Cell Mol. Biol.* **348**, 69–121 (2019).
42. Neefjes, J., Jongsma, M. L. M., Paul, P. & Bakke, O. Towards a systems understanding of MHC class I and MHC class II antigen presentation. *Nat. Rev. Immunol.* **11**, 823–836 (2011).
43. Rock, K. L., Reits, E. & Neefjes, J. Present Yourself! By MHC Class I and MHC Class II Molecules. *Trends Immunol.* **37**, 724–737 (2016).
44. Raphael, I., Joern, R. R. & Forsthuber, T. G. Memory CD4+ T Cells in Immunity and Autoimmune Diseases. *Cells* **9**, (2020).
45. Sprent, J. & Surh, C. D. T Cell Memory. *Annu. Rev. Immunol.* **20**, 551–579 (2002).
46. Jenkins, M. K. *et al.* In Vivo Activation of Antigen-Specific CD4 T Cells. *Annu. Rev. Immunol.* **19**, 23–45 (2001).
47. McHeyzer-Williams, L. J. & McHeyzer-Williams, M. G. Antigen-specific memory b cell development. *Annu. Rev. Immunol.* **23**, 487–513 (2004).
48. Weisel, F. & Shlomchik, M. Memory B Cells of Mice and Humans. *Annu. Rev. Immunol.* **35**, 255–284 (2017).
49. Charles A Janeway, J., Travers, P., Walport, M. & Shlomchik, M. J. The distribution and functions of immunoglobulin isotypes. *Immunobiol. Immune Syst. Health Dis. 5th Ed.* (2001).
50. Schroeder, H. W. & Cavacini, L. Structure and Function of Immunoglobulins. *J. Allergy Clin. Immunol.* **125**, S41–S52 (2010).
51. Gül, N. & Egmond, M. van. Antibody-Dependent Phagocytosis of Tumor Cells by Macrophages: A Potent Effector Mechanism of Monoclonal Antibody Therapy of Cancer. *Cancer Res.* **75**, 5008–5013 (2015).
52. Golstein, P. & Griffiths, G. M. An early history of T cell-mediated cytotoxicity. *Nat. Rev. Immunol.* **18**, 527–535 (2018).
53. Zhang, N. & Bevan, M. J. CD8+ T Cells: Foot Soldiers of the Immune System. *Immunity* **35**, 161–168 (2011).
54. Zhu, J., Yamane, H. & Paul, W. E. Differentiation of Effector CD4 T Cell Populations. *Annu. Rev. Immunol.* **28**, 445–489 (2010).
55. Ruterbusch, M., Pruner, K. B., Shehata, L. & Pepper, M. In Vivo CD4+ T Cell Differentiation and Function: Revisiting the Th1/Th2 Paradigm. *Annu. Rev. Immunol.* **38**, 705–725 (2020).
56. Welsh, R. M., Selin, L. K. & Szomolanyi-Tsuda, E. Immunological Memory to Viral Infections. *Annu. Rev. Immunol.* **22**, 711–743 (2004).
57. Zinkernagel, R. M. On Natural and Artificial Vaccinations. *Annu. Rev. Immunol.* **21**, 515–546 (2003).
58. Godfrey, D. I., Uldrich, A. P., McCluskey, J., Rossjohn, J. & Moody, D. B. The burgeoning family of unconventional T cells. *Nat. Immunol.* **16**, 1114–1123 (2015).
59. Taniuchi, I. CD4 Helper and CD8 Cytotoxic T Cell Differentiation. *Annu. Rev. Immunol.* **36**, 579–601 (2018).
60. Kondo, K., Ohigashi, I. & Takahama, Y. Thymus machinery for T-cell selection. *Int. Immunol.* **31**, 119–125 (2018).
61. Thapa, P. & Farber, D. L. The role of the thymus in the immune response. *Thorac. Surg. Clin.* **29**, 123–131 (2019).
62. Perniola, R. Twenty Years of AIRE. *Front. Immunol.* **9**, (2018).
63. Passos, G. A., Speck-Hernandez, C. A., Assis, A. F. & Mendes-da-Cruz, D. A. Update on Aire and thymic negative selection. *Immunology* **153**, 10–20 (2018).
64. Stritesky, G. L., Jameson, S. C. & Hogquist, K. A. Selection of Self-Reactive T Cells in the Thymus. *Annu. Rev. Immunol.* **30**, 95–114 (2012).
65. Smith-Garvin, J. E., Koretzky, G. A. & Jordan, M. S. T Cell Activation. *Annu. Rev. Immunol.* **27**, 591–619 (2009).
66. Jenkins, M. K. & Schwartz, R. H. Antigen presentation by chemically modified splenocytes induces antigen-specific T cell unresponsiveness in vitro and in vivo. *J. Exp. Med.* **165**, 302–319 (1987).
67. Schwartz, R. H. T Cell Anergy. *Annu. Rev. Immunol.* **21**, 305–334 (2003).
68. Chen, L. & Flies, D. B. Molecular mechanisms of T cell co-stimulation and co-inhibition. *Nat. Rev. Immunol.* **13**, 227–242 (2013).
69. Nagai, S. & Azuma, M. The CD28-B7 Family of Co-signaling Molecules. *Adv. Exp. Med. Biol.* **1189**, 25–51 (2019).
70. Dos Santos Meira, C. & Gedamu, L. Protective or Detrimental? Understanding the Role of Host Immunity in Leishmaniasis. *Microorganisms* **7**, (2019).
71. de Sousa, J. R., Sotto, M. N. & Simões Quaresma, J. A. Leprosy As a Complex Infection: Breakdown of the Th1 and Th2 Immune Paradigm in the Immunopathogenesis of the Disease. *Front. Immunol.* **8**, (2017).

72. Brucklacher-Waldert, V., Carr, E. J., Linterman, M. A. & Veldhoen, M. Cellular Plasticity of CD4+ T Cells in the Intestine. *Front. Immunol.* **5**, (2014).
73. DuPage, M. & Bluestone, J. A. Harnessing the plasticity of CD4+ T cells to treat immune-mediated disease. *Nat. Rev. Immunol.* **16**, 149–163 (2016).
74. O'Shea, J. J. & Paul, W. E. Mechanisms Underlying Lineage Commitment and Plasticity of Helper CD4+ T Cells. *Science* **327**, 1098–1102 (2010).
75. Romagnani, S. The Th1/Th2 paradigm. *Immunol. Today* **18**, 263–266 (1997).
76. Iwamoto, I., Nakajima, H., Endo, H. & Yoshida, S. Interferon gamma regulates antigen-induced eosinophil recruitment into the mouse airways by inhibiting the infiltration of CD4+ T cells. *J. Exp. Med.* **177**, 573–576 (1993).
77. Cohn, L., Homer, R. J., Niu, N. & Bottomly, K. T Helper 1 Cells and Interferon γ Regulate Allergic Airway Inflammation and Mucus Production. *J. Exp. Med.* **190**, 1309–1318 (1999).
78. Dow, S. W., Schwarze, J., Heath, T. D., Potter, T. A. & Gelfand, E. W. Systemic and local interferon gamma gene delivery to the lungs for treatment of allergen-induced airway hyperresponsiveness in mice. *Hum. Gene Ther.* **10**, 1905–1914 (1999).
79. Wei, G. *et al.* Global Mapping of H3K4me3 and H3K27me3 Reveals Specificity and Plasticity in Lineage Fate Determination of Differentiating CD4+ T Cells. *Immunity* **30**, 155–167 (2009).
80. Szabo, S. J. *et al.* A novel transcription factor, T-bet, directs Th1 lineage commitment. *Cell* **100**, 655–669 (2000).
81. Mullen, A. C. *et al.* Role of T-bet in commitment of TH1 cells before IL-12-dependent selection. *Science* **292**, 1907–1910 (2001).
82. Kohu, K. *et al.* The Runx3 transcription factor augments Th1 and down-modulates Th2 phenotypes by interacting with and attenuating GATA3. *J. Immunol. Baltim. Md 1950* **183**, 7817–7824 (2009).
83. Zheng, W. *et al.* Up-regulation of Hlx in immature Th cells induces IFN-gamma expression. *J. Immunol. Baltim. Md 1950* **172**, 114–122 (2004).
84. Szabo, S. J., Sullivan, B. M., Peng, S. L. & Glimcher, L. H. Molecular Mechanisms RegulatinG Th1 Immune Responses. *Annu. Rev. Immunol.* **21**, 713–758 (2003).
85. Groom, J. R. & Luster, A. D. CXCR3 in T cell function. *Exp. Cell Res.* **317**, 620–631 (2011).
86. Qin, S. *et al.* The chemokine receptors CXCR3 and CCR5 mark subsets of T cells associated with certain inflammatory reactions. *J. Clin. Invest.* **101**, 746–754 (1998).
87. Xie, J. H. *et al.* Antibody-mediated blockade of the CXCR3 chemokine receptor results in diminished recruitment of T helper 1 cells into sites of inflammation. *J. Leukoc. Biol.* **73**, 771–780 (2003).
88. Germann, T. *et al.* Interleukin-12 profoundly up-regulates the synthesis of antigen-specific complement-fixing IgG2a, IgG2b and IgG3 antibody subclasses in vivo. *Eur. J. Immunol.* **25**, 823–829 (1995).
89. Stevens, T. L. *et al.* Regulation of antibody isotype secretion by subsets of antigen-specific helper T cells. *Nature* **334**, 255–258 (1988).
90. Chapman, S. J. & Hill, A. V. S. Human genetic susceptibility to infectious disease. *Nat. Rev. Genet.* **13**, 175–188 (2012).
91. de Jong, R. *et al.* Severe mycobacterial and Salmonella infections in interleukin-12 receptor-deficient patients. *Science* **280**, 1435–1438 (1998).
92. Jouanguy, E. *et al.* Interferon-γ –Receptor Deficiency in an Infant with Fatal Bacille Calmette–Guérin Infection. *N. Engl. J. Med.* **335**, 1956–1962 (1996).
93. Walker, J. A. & McKenzie, A. N. J. T H 2 cell development and function. *Nat. Rev. Immunol.* **18**, 121–133 (2018).
94. Nakayama, T. *et al.* Th2 Cells in Health and Disease. *Annu. Rev. Immunol.* **35**, 53–84 (2017).
95. Gutierrez, D. A. & Rodewald, H.-R. A Sting in the Tale of Th2 Immunity. *Immunity* **39**, 803–805 (2013).
96. Yamane, H. & Paul, W. E. Early signaling events that underlie fate decisions of naive CD4+ T cells towards distinct T-helper cell subsets. *Immunol. Rev.* **252**, 12–23 (2013).
97. Omori, M. & Ziegler, S. Induction of IL-4 Expression in CD4+ T Cells by Thymic Stromal Lymphopoietin. *J. Immunol.* **178**, 1396–1404 (2007).
98. Rochman, I., Watanabe, N., Arima, K., Liu, Y.-J. & Leonard, W. J. Cutting Edge: Direct Action of Thymic Stromal Lymphopoietin on Activated Human CD4+ T Cells. *J. Immunol.* **178**, 6720–6724 (2007).
99. Turner, H. & Kinet, J.-P. Signalling through the high-affinity IgE receptor FcεRI. *Nature* **402**, 24–30 (1999).
100. Mikhak, Z. *et al.* Contribution of CCR4 and CCR8 to antigen-specific Th2 cell trafficking in allergic pulmonary inflammation. *J. Allergy Clin. Immunol.* **123**, 67-73.e3 (2009).
101. Harrington, L. E. *et al.* Interleukin 17-producing CD4+ effector T cells develop via a lineage distinct from the T helper type 1 and 2 lineages. *Nat. Immunol.* **6**, 1123–1132 (2005).
102. Korn, T., Bettelli, E., Oukka, M. & Kuchroo, V. K. IL-17 and Th17 Cells. *Annu. Rev. Immunol.* **27**, 485–517 (2009).
103. Mulcahy, M. E., Leech, J. M., Renauld, J.-C., Mills, K. H. & McLoughlin, R. M. Interleukin-22 regulates antimicrobial peptide expression and keratinocyte differentiation to control Staphylococcus aureus colonization of the nasal mucosa. *Mucosal Immunol.* **9**, 1429–1441 (2016).
104. Sugimoto, K. *et al.* IL-22 ameliorates intestinal inflammation in a mouse model of ulcerative colitis. *J. Clin. Invest.* **118**, 534–544 (2008).
105. Ivanov, I. I. *et al.* The Orphan Nuclear Receptor RORγt Directs the Differentiation Program of Proinflammatory IL-17+ T Helper Cells. *Cell* **126**, 1121–1133 (2006).

106. Yasuda, K., Takeuchi, Y. & Hirota, K. The pathogenicity of Th17 cells in autoimmune diseases. *Semin. Immunopathol.* **41**, 283–297 (2019).

107. Yeh, W.-I., McWilliams, I. L. & Harrington, L. E. IFNγ inhibits Th17 differentiation and function via Tbet-dependent and Tbet-independent mechanisms. *J. Neuroimmunol.* **267**, 20–27 (2014).

108. Diveu, C. *et al.* IL-27 blocks RORc expression to inhibit lineage commitment of Th17 cells. *J. Immunol. Baltim. Md 1950* **182**, 5748–5756 (2009).

109. Stumhofer, J. S. *et al.* Interleukin 27 negatively regulates the development of interleukin 17-producing T helper cells during chronic inflammation of the central nervous system. *Nat. Immunol.* **7**, 937–945 (2006).

110. Stritesky, G. L., Yeh, N. & Kaplan, M. H. IL-23 promotes maintenance but not commitment to the Th17 lineage. *J. Immunol. Baltim. Md 1950* **181**, 5948–5955 (2008).

111. El-Behi, M. *et al.* The encephalitogenicity of TH17 cells is dependent on IL-1- and IL-23-induced production of the cytokine GM-CSF. *Nat. Immunol.* **12**, 568–575 (2011).

112. Ghoreschi, K. *et al.* Generation of Pathogenic Th17 Cells in the Absence of TGF-β Signaling. *Nature* **467**, 967–971 (2010).

113. Kamali, A. N. *et al.* A role for Th1-like Th17 cells in the pathogenesis of inflammatory and autoimmune disorders. *Mol. Immunol.* **105**, 107–115 (2019).

114. Lee, Y. *et al.* Induction and molecular signature of pathogenic T H 17 cells. *Nat. Immunol.* **13**, 991–999 (2012).

115. Chen, Z. *et al.* FOXP3 and RORγt: Transcriptional regulation of Treg and Th17. *Int. Immunopharmacol.* **11**, 536–542 (2011).

116. Chen, X. & Oppenheim, J. J. Th17 cells and Tregs: unlikely allies. *J. Leukoc. Biol.* **95**, 723–731 (2014).

117. Kimura, A. & Kishimoto, T. IL-6: Regulator of Treg/Th17 balance. *Eur. J. Immunol.* **40**, 1830–1835 (2010).

118. Gagliani, N. *et al.* Th17 cells transdifferentiate into regulatory T cells during resolution of inflammation. *Nature* **523**, 221–225 (2015).

119. Breitfeld, D. *et al.* Follicular B Helper T Cells Express Cxc Chemokine Receptor 5, Localize to B Cell Follicles, and Support Immunoglobulin Production. *J. Exp. Med.* **192**, 1545–1552 (2000).

120. Schaerli, P. *et al.* Cxc Chemokine Receptor 5 Expression Defines Follicular Homing T Cells with B Cell Helper Function. *J. Exp. Med.* **192**, 1553–1562 (2000).

121. Kim, C. H. *et al.* Subspecialization of Cxcr5+ T Cells. *J. Exp. Med.* **193**, 1373–1382 (2001).

122. Crotty, S. Follicular Helper CD4 T Cells (TFH). *Annu. Rev. Immunol.* **29**, 621–663 (2011).

123. Vinuesa, C. G., Linterman, M. A., Yu, D. & MacLennan, I. C. M. Follicular Helper T Cells. *Annu. Rev. Immunol.* **34**, 335–368 (2016).

124. Nurieva, R. I. *et al.* Generation of T Follicular Helper Cells Is Mediated by Interleukin-21 but Independent of T Helper 1, 2, or 17 Cell Lineages. *Immunity* **29**, 138–149 (2008).

125. Eddahri, F. *et al.* Interleukin-6/STAT3 signaling regulates the ability of naive T cells to acquire B-cell help capacities. *Blood* **113**, 2426–2433 (2009).

126. Chtanova, T. *et al.* T Follicular Helper Cells Express a Distinctive Transcriptional Profile, Reflecting Their Role as Non-Th1/Th2 Effector Cells That Provide Help for B Cells. *J. Immunol.* **173**, 68–78 (2004).

127. Bauquet, A. T. *et al.* The costimulatory molecule ICOS regulates the expression of c-Maf and IL-21 in the development of follicular T helper cells and TH-17 cells. *Nat. Immunol.* **10**, 167–175 (2009).

128. Liu, X. *et al.* Bcl6 expression specifies the T follicular helper cell program in vivo. *J. Exp. Med.* **209**, 1841–1852 (2012).

129. Kroenke, M. A. *et al.* Bcl6 and Maf Cooperate To Instruct Human Follicular Helper CD4 T Cell Differentiation. *J. Immunol.* **188**, 3734–3744 (2012).

130. Reinhardt, R. L., Liang, H.-E. & Locksley, R. M. Cytokine-secreting follicular T cells shape the antibody repertoire. *Nat. Immunol.* **10**, 385–393 (2009).

131. Koch, S., Sopel, N. & Finotto, S. Th9 and other IL-9-producing cells in allergic asthma. *Semin. Immunopathol.* **39**, 55–68 (2017).

132. Louahed, J. *et al.* Interleukin-9 upregulates mucus expression in the airways. *Am. J. Respir. Cell Mol. Biol.* **22**, 649–656 (2000).

133. Hültner, L. *et al.* Mast cell growth-enhancing activity (MEA) is structurally related and functionally identical to the novel mouse T cell growth factor P40/TCGFIII (interleukin 9). *Eur. J. Immunol.* **20**, 1413–1416 (1990).

134. Veldhoen, M. *et al.* Transforming growth factor-beta 'reprograms' the differentiation of T helper 2 cells and promotes an interleukin 9-producing subset. *Nat. Immunol.* **9**, 1341–1346 (2008).

135. Chen, T. *et al.* Th9 Cell Differentiation and Its Dual Effects in Tumor Development. *Front. Immunol.* **11**, (2020).

136. Duhen, T., Geiger, R., Jarrossay, D., Lanzavecchia, A. & Sallusto, F. Production of interleukin 22 but not interleukin 17 by a subset of human skin-homing memory T cells. *Nat. Immunol.* **10**, 857–863 (2009).

137. Eyerich, S. *et al.* Th22 cells represent a distinct human T cell subset involved in epidermal immunity and remodeling. *J. Clin. Invest.* **119**, 3573–3585 (2009).

138. Fujita, H. *et al.* Human Langerhans cells induce distinct IL-22-producing CD4+ T cells lacking IL-17 production. *Proc. Natl. Acad. Sci. U. S. A.* **106**, 21795–21800 (2009).

139. Trifari, S., Kaplan, C. D., Tran, E. H., Crellin, N. K. & Spits, H. Identification of a human helper T cell population that has abundant production of interleukin 22 and is distinct from T(H)-17, T(H)1 and T(H)2 cells. *Nat. Immunol.* **10**, 864–871 (2009).
140. Monteleone, I. *et al.* Aryl hydrocarbon receptor-induced signals up-regulate IL-22 production and inhibit inflammation in the gastrointestinal tract. *Gastroenterology* **141**, 237–248, 248.e1 (2011).
141. Azizi, G., Yazdani, R. & Mirshafiey, A. Th22 cells in autoimmunity: a review of current knowledge. *Eur. Ann. Allergy Clin. Immunol.* **47**, 108–117 (2015).
142. Grivennikov, S. I. *et al.* Adenoma-linked barrier defects and microbial products drive IL-23/IL-17-mediated tumour growth. *Nature* **491**, 254–258 (2012).
143. Zheng, Y. *et al.* Interleukin-22, a T(H)17 cytokine, mediates IL-23-induced dermal inflammation and acanthosis. *Nature* **445**, 648–651 (2007).
144. Pereira, L. M. S., Gomes, S. T. M., Ishak, R. & Vallinoto, A. C. R. Regulatory T Cell and Forkhead Box Protein 3 as Modulators of Immune Homeostasis. *Front. Immunol.* **8**, (2017).
145. Gershon, R. K. & Kondo, K. Cell interactions in the induction of tolerance: the role of thymic lymphocytes. *Immunology* **18**, 723–737 (1970).
146. Sakaguchi, S., Wing, K. & Miyara, M. Regulatory T cells – a brief history and perspective. *Eur. J. Immunol.* **37**, S116–S123 (2007).
147. Sakaguchi, S., Sakaguchi, N., Asano, M., Itoh, M. & Toda, M. Immunologic self-tolerance maintained by activated T cells expressing IL-2 receptor alpha-chains (CD25). Breakdown of a single mechanism of self-tolerance causes various autoimmune diseases. *J. Immunol. Baltim. Md 1950* **155**, 1151–1164 (1995).
148. Asano, M., Toda, M., Sakaguchi, N. & Sakaguchi, S. Autoimmune disease as a consequence of developmental abnormality of a T cell subpopulation. *J. Exp. Med.* **184**, 387–396 (1996).
149. Khattri, R., Cox, T., Yasayko, S.-A. & Ramsdell, F. An essential role for Scurfin in CD4 + CD25 + T regulatory cells. *Nat. Immunol.* **4**, 337–342 (2003).
150. Hori, S., Nomura, T. & Sakaguchi, S. Control of Regulatory T Cell Development by the Transcription Factor Foxp3. *Science* **299**, 1057–1061 (2003).
151. Fontenot, J. D., Gavin, M. A. & Rudensky, A. Y. Foxp3 programs the development and function of CD4+CD25+ regulatory T cells. *Nat. Immunol.* **4**, 330–336 (2003).
152. Lu, L., Barbi, J. & Pan, F. The regulation of immune tolerance by FOXP3. *Nat. Rev. Immunol.* **17**, 703–717 (2017).
153. Brunkow, M. E. *et al.* Disruption of a new forkhead/winged-helix protein, scurfin, results in the fatal lymphoproliferative disorder of the scurfy mouse. *Nat. Genet.* **27**, 68–73 (2001).
154. Wildin, R. S. *et al.* X-linked neonatal diabetes mellitus, enteropathy and endocrinopathy syndrome is the human equivalent of mouse scurfy. *Nat. Genet.* **27**, 18–20 (2001).
155. Bennett, C. L. *et al.* The immune dysregulation, polyendocrinopathy, enteropathy, X-linked syndrome (IPEX) is caused by mutations of FOXP3. *Nat. Genet.* **27**, 20–21 (2001).
156. Rodríguez-Perea, A. L., Arcia, E. D., Rueda, C. M. & Velilla, P. A. Phenotypical characterization of regulatory T cells in humans and rodents. *Clin. Exp. Immunol.* **185**, 281–291 (2016).
157. Foxp3 - Forkhead box protein P3 - Mus musculus (Mouse) - Foxp3 gene & protein. https://www.uniprot.org/uniprot/Q99JB6.
158. FOXP3 - Forkhead box protein P3 - Homo sapiens (Human) - FOXP3 gene & protein. https://www.uniprot.org/uniprot/Q9BZS1.
159. Kaufmann, E. & Knöchel, W. Five years on the wings of fork head. *Mech. Dev.* **57**, 3–20 (1996).
160. Xie, X. *et al.* The Regulatory T Cell Lineage Factor Foxp3 Regulates Gene Expression through Several Distinct Mechanisms Mostly Independent of Direct DNA Binding. *PLoS Genet.* **11**, e1005251 (2015).
161. Gambineri, E., Torgerson, T. R. & Ochs, H. D. Immune dysregulation, polyendocrinopathy, enteropathy, and X-linked inheritance (IPEX), a syndrome of systemic autoimmunity caused by mutations of FOXP3, a critical regulator of T-cell homeostasis. *Curr. Opin. Rheumatol.* **15**, 430–435 (2003).
162. Lopes, J. E. *et al.* Analysis of FOXP3 Reveals Multiple Domains Required for Its Function as a Transcriptional Repressor. *J. Immunol.* **177**, 3133–3142 (2006).
163. Li, B. *et al.* FOXP3 is a homo-oligomer and a component of a supramolecular regulatory complex disabled in the human XLAAD/IPEX autoimmune disease. *Int. Immunol.* **19**, 825–835 (2007).
164. Deng, G. *et al.* Molecular and biological role of the FOXP3 N-terminal domain in immune regulation by T regulatory/suppressor cells. *Exp. Mol. Pathol.* **93**, 334–338 (2012).
165. Rudra, D. *et al.* Transcription factor Foxp3 and its protein partners form a complex regulatory network. *Nat. Immunol.* **13**, 1010–1019 (2012).
166. Li, B. *et al.* FOXP3 interactions with histone acetyltransferase and class II histone deacetylases are required for repression. *Proc. Natl. Acad. Sci. U. S. A.* **104**, 4571–4576 (2007).
167. Zheng, Y. *et al.* Genome-wide analysis of Foxp3 target genes in developing and mature regulatory T cells. *Nature* **445**, 936–940 (2007).
168. Marson, A. *et al.* Foxp3 occupancy and regulation of key target genes during T-cell stimulation. *Nature* **445**, 931–935 (2007).

169. Fu, W. *et al.* A multiply redundant genetic switch 'locks in' the transcriptional signature of regulatory T cells. *Nat. Immunol.* **13**, 972–980 (2012).
170. Hori, S. The Foxp3 interactome: a network perspective of T reg cells. *Nat. Immunol.* **13**, 943–945 (2012).
171. Wu, Y. *et al.* FOXP3 controls regulatory T cell function through cooperation with NFAT. *Cell* **126**, 375–387 (2006).
172. Ono, M. *et al.* Foxp3 controls regulatory T-cell function by interacting with AML1/Runx1. *Nature* **446**, 685–689 (2007).
173. Bettelli, E., Dastrange, M. & Oukka, M. Foxp3 interacts with nuclear factor of activated T cells and NF-κB to repress cytokine gene expression and effector functions of T helper cells. *Proc. Natl. Acad. Sci. U. S. A.* **102**, 5138–5143 (2005).
174. Josefowicz, S. Z., Lu, L.-F. & Rudensky, A. Y. Regulatory T Cells: Mechanisms of Differentiation and Function. *Annu. Rev. Immunol.* **30**, 531–564 (2012).
175. Ouyang, W. *et al.* Foxo proteins cooperatively control the differentiation of Foxp3 + regulatory T cells. *Nat. Immunol.* **11**, 618–627 (2010).
176. Kim, H.-P. & Leonard, W. J. CREB/ATF-dependent T cell receptor–induced FoxP3 gene expression: a role for DNA methylation. *J. Exp. Med.* **204**, 1543–1551 (2007).
177. Zheng, Y. *et al.* Role of conserved non-coding DNA elements in the Foxp3 gene in regulatory T-cell fate. *Nature* **463**, 808–812 (2010).
178. Kitagawa, Y. *et al.* Guidance of regulatory T cell development by Satb1-dependent super-enhancer establishment. *Nat. Immunol.* **18**, 173–183 (2017).
179. Abbas, A. K. *et al.* Regulatory T cells: recommendations to simplify the nomenclature. *Nat. Immunol.* **14**, 307–308 (2013).
180. Moran, A. E. *et al.* T cell receptor signal strength in Treg and iNKT cell development demonstrated by a novel fluorescent reporter mouse. *J. Exp. Med.* **208**, 1279–1289 (2011).
181. Au-Yeung, B. B. *et al.* A sharp T-cell antigen receptor signaling threshold for T-cell proliferation. *Proc. Natl. Acad. Sci.* **111**, E3679–E3688 (2014).
182. Fassett, M. S., Jiang, W., D'Alise, A. M., Mathis, D. & Benoist, C. Nuclear receptor Nr4a1 modulates both regulatory T-cell (Treg) differentiation and clonal deletion. *Proc. Natl. Acad. Sci. U. S. A.* **109**, 3891–3896 (2012).
183. Hori, S. c-Rel: A pioneer in directing regulatory T-cell lineage commitment? *Eur. J. Immunol.* **40**, 664–667 (2010).
184. Long, M., Park, S.-G., Strickland, I., Hayden, M. S. & Ghosh, S. Nuclear Factor-κB Modulates Regulatory T Cell Development by Directly Regulating Expression of Foxp3 Transcription Factor. *Immunity* **31**, 921–931 (2009).
185. Kitagawa, Y., Ohkura, N. & Sakaguchi, S. Molecular Determinants of Regulatory T Cell Development: The Essential Roles of Epigenetic Changes. *Front. Immunol.* **4**, (2013).
186. Burchill, M. A., Yang, J., Vogtenhuber, C., Blazar, B. R. & Farrar, M. A. IL-2 Receptor β-Dependent STAT5 Activation Is Required for the Development of Foxp3+ Regulatory T Cells. *J. Immunol.* **178**, 280–290 (2007).
187. Yao, Z. *et al.* Nonredundant roles for Stat5a/b in directly regulating Foxp3. *Blood* **109**, 4368–4375 (2007).
188. Laurence, A. *et al.* STAT3 Transcription Factor Promotes Instability of nTreg Cells and Limits Generation of iTreg Cells During Acute Murine Graft Versus Host Disease. *Immunity* **37**, 209–222 (2012).
189. Tanoue, T., Atarashi, K. & Honda, K. Development and maintenance of intestinal regulatory T cells. *Nat. Rev. Immunol.* **16**, 295–309 (2016).
190. Tone, Y. *et al.* Smad3 and NFAT cooperate to induce Foxp3 expression through its enhancer. *Nat. Immunol.* **9**, 194–202 (2008).
191. Elias, K. M. *et al.* Retinoic acid inhibits Th17 polarization and enhances FoxP3 expression through a Stat-3/Stat-5 independent signaling pathway. *Blood* **111**, 1013–1020 (2008).
192. Xu, L. *et al.* Positive and Negative Transcriptional Regulation of the Foxp3 gene is Mediated by TGF-β Signal Transducer Smad3 Access and Binding to Enhancer I. *Immunity* **33**, 313–325 (2010).
193. Xiao, S. *et al.* Retinoic acid increases Foxp3+ regulatory T cells and inhibits development of Th17 cells by enhancing TGF-β-driven Smad3 signaling and inhibiting IL-6 and IL-23 receptor expression. *J. Immunol. Baltim. Md 1950* **181**, 2277–2284 (2008).
194. Lu, L. *et al.* All-Trans Retinoic Acid Promotes TGF-β-Induced Tregs via Histone Modification but Not DNA Demethylation on Foxp3 Gene Locus. *PLoS ONE* **6**, (2011).
195. Arpaia, N. *et al.* Metabolites produced by commensal bacteria promote peripheral regulatory T-cell generation. *Nature* **504**, 451–455 (2013).
196. Furusawa, Y. *et al.* Commensal microbe-derived butyrate induces the differentiation of colonic regulatory T cells. *Nature* **504**, 446–450 (2013).
197. Zheng, S. G., Wang, J., Wang, P., Gray, J. D. & Horwitz, D. A. IL-2 Is Essential for TGF-β to Convert Naive CD4+CD25− Cells to CD25+Foxp3+ Regulatory T Cells and for Expansion of These Cells. *J. Immunol.* **178**, 2018–2027 (2007).
198. Miyao, T. *et al.* Plasticity of Foxp3+ T Cells Reflects Promiscuous Foxp3 Expression in Conventional T Cells but Not Reprogramming of Regulatory T Cells. *Immunity* **36**, 262–275 (2012).
199. Ohkura, N. *et al.* T Cell Receptor Stimulation-Induced Epigenetic Changes and Foxp3 Expression Are Independent and Complementary Events Required for Treg Cell Development. *Immunity* **37**, 785–799 (2012).

200. Hilbrands, R. *et al.* Induced Foxp3+ T Cells Colonizing Tolerated Allografts Exhibit the Hypomethylation Pattern Typical of Mature Regulatory T Cells. *Front. Immunol.* **7**, (2016).
201. Thornton, A. M. *et al.* Expression of Helios, an Ikaros Transcription Factor Family Member, Differentiates Thymic-Derived from Peripherally Induced Foxp3+ T Regulatory Cells. *J. Immunol.* **184**, 3433–3441 (2010).
202. Weiss, J. M. *et al.* Neuropilin 1 is expressed on thymus-derived natural regulatory T cells, but not mucosa-generated induced Foxp3+ T reg cells. *J. Exp. Med.* **209**, 1723–1742 (2012).
203. Yadav, M. *et al.* Neuropilin-1 distinguishes natural and inducible regulatory T cells among regulatory T cell subsets in vivo. *J. Exp. Med.* **209**, 1713–1722 (2012).
204. Yadav, M., Bluestone, J. A. & Stephan, S. Peripherally Induced Tregs – Role in Immune Homeostasis and Autoimmunity. *Front. Immunol.* **4**, (2013).
205. Thornton, A. M. *et al.* Helios+ and Helios− Treg subpopulations are phenotypically and functionally distinct and express dissimilar TCR repertoires. *Eur. J. Immunol.* **49**, 398–412 (2019).
206. Pacholczyk, R., Ignatowicz, H., Kraj, P. & Ignatowicz, L. Origin and T Cell Receptor Diversity of Foxp3+CD4+CD25+ T Cells. *Immunity* **25**, 249–259 (2006).
207. Hsieh, C.-S., Zheng, Y., Liang, Y., Fontenot, J. D. & Rudensky, A. Y. An intersection between the self-reactive regulatory and nonregulatory T cell receptor repertoires. *Nat. Immunol.* **7**, 401–410 (2006).
208. Wong, J., Mathis, D. & Benoist, C. TCR-based lineage tracing: no evidence for conversion of conventional into regulatory T cells in response to a natural self-antigen in pancreatic islets. *J. Exp. Med.* **204**, 2039–2045 (2007).
209. Thornton, A. M. & Shevach, E. M. Suppressor effector function of CD4+CD25+ immunoregulatory T cells is antigen nonspecific. *J. Immunol. Baltim. Md 1950* **164**, 183–190 (2000).
210. Karim, M., Feng, G., Wood, K. J. & Bushell, A. R. CD25+CD4+ regulatory T cells generated by exposure to a model protein antigen prevent allograft rejection: antigen-specific reactivation in vivo is critical for bystander regulation. *Blood* **105**, 4871–4877 (2005).
211. Shevyrev, D. & Tereshchenko, V. Treg Heterogeneity, Function, and Homeostasis. *Front. Immunol.* **10**, (2020).
212. Travis, M. A. & Sheppard, D. TGF-β Activation and Function in Immunity. *Annu. Rev. Immunol.* **32**, 51–82 (2014).
213. Shi, M. *et al.* Latent TGF-β structure and activation. *Nature* **474**, 343–349 (2011).
214. Annes, J. P., Chen, Y., Munger, J. S. & Rifkin, D. B. Integrin αVβ6-mediated activation of latent TGF-β requires the latent TGF-β binding protein-1. *J. Cell Biol.* **165**, 723–734 (2004).
215. Liénart, S. *et al.* Structural basis of latent TGF-β1 presentation and activation by GARP on human regulatory T cells. *Science* **362**, 952–956 (2018).
216. Glinka, Y. & Prud'homme, G. J. Neuropilin-1 is a receptor for transforming growth factor β-1, activates its latent form, and promotes regulatory T cell activity. *J. Leukoc. Biol.* **84**, 302–310 (2008).
217. Rubtsov, Y. P. *et al.* Regulatory T Cell-Derived Interleukin-10 Limits Inflammation at Environmental Interfaces. *Immunity* **28**, 546–558 (2008).
218. Chaudhry, A. *et al.* Interleukin-10 signaling in regulatory T cells is required for suppression of Th17 cell-mediated inflammation. *Immunity* **34**, 566–578 (2011).
219. Huber, S. *et al.* Th17 cells express interleukin-10 receptor and are controlled by Foxp3− and Foxp3+ regulatory CD4+ T cells in an interleukin-10 dependent manner. *Immunity* **34**, 554–565 (2011).
220. Collison, L. W. *et al.* The inhibitory cytokine IL-35 contributes to regulatory T-cell function. *Nature* **450**, 566–569 (2007).
221. Sawant, D. V. *et al.* Adaptive plasticity of IL-10 + and IL-35 + T reg cells cooperatively promotes tumor T cell exhaustion. *Nat. Immunol.* **20**, 724–735 (2019).
222. Thornton, A. M. & Shevach, E. M. CD4+CD25+ Immunoregulatory T Cells Suppress Polyclonal T Cell Activation In Vitro by Inhibiting Interleukin 2 Production. *J. Exp. Med.* **188**, 287–296 (1998).
223. Rosa, M. de la, Rutz, S., Dorninger, H. & Scheffold, A. Interleukin-2 is essential for CD4+CD25+ regulatory T cell function. *Eur. J. Immunol.* **34**, 2480–2488 (2004).
224. Pandiyan, P., Zheng, L., Ishihara, S., Reed, J. & Lenardo, M. J. CD4 + CD25 + Foxp3 + regulatory T cells induce cytokine deprivation–mediated apoptosis of effector CD4 + T cells. *Nat. Immunol.* **8**, 1353–1362 (2007).
225. Höfer, T., Krichevsky, O. & Altan-Bonnet, G. Competition for IL-2 between Regulatory and Effector T Cells to Chisel Immune Responses. *Front. Immunol.* **3**, (2012).
226. McNally, A., Hill, G. R., Sparwasser, T., Thomas, R. & Steptoe, R. J. CD4+CD25+ regulatory T cells control CD8+ T-cell effector differentiation by modulating IL-2 homeostasis. *Proc. Natl. Acad. Sci.* **108**, 7529–7534 (2011).
227. Voisinne, G. *et al.* "T cells integrate Local and Global cues to discriminate between structurally similar antigens". *Cell Rep.* **11**, 1208–1219 (2015).
228. O'Brien, S. *et al.* Ikaros imposes a barrier to CD8+ T cell differentiation by restricting autocrine IL-2 production. *J. Immunol. Baltim. Md 1950* **192**, 5118–5129 (2014).
229. Deaglio, S. *et al.* Adenosine generation catalyzed by CD39 and CD73 expressed on regulatory T cells mediates immune suppression. *J. Exp. Med.* **204**, 1257–1265 (2007).
230. Cekic, C. & Linden, J. Purinergic regulation of the immune system. *Nat. Rev. Immunol.* **16**, 177–192 (2016).
231. Allard, D., Turcotte, M. & Stagg, J. Targeting A2 adenosine receptors in cancer. *Immunol. Cell Biol.* **95**, 333–339 (2017).

232. Bopp, T. *et al.* Cyclic adenosine monophosphate is a key component of regulatory T cell–mediated suppression. *J. Exp. Med.* **204**, 1303–1310 (2007).

233. Klein, M. & Bopp, T. Cyclic AMP Represents a Crucial Component of Treg Cell-Mediated Immune Regulation. *Front. Immunol.* **7**, (2016).

234. Sansom, D. M. CD28, CTLA-4 and their ligands: who does what and to whom? *Immunology* **101**, 169–177 (2000).

235. Onishi, Y., Fehervari, Z., Yamaguchi, T. & Sakaguchi, S. Foxp3+ natural regulatory T cells preferentially form aggregates on dendritic cells in vitro and actively inhibit their maturation. *Proc. Natl. Acad. Sci.* **105**, 10113–10118 (2008).

236. Akkaya, B. *et al.* Ex-vivo iTreg differentiation revisited: Convenient alternatives to existing strategies. *J. Immunol. Methods* **441**, 67–71 (2017).

237. Akkaya, B. *et al.* Tregs orchestrate antigen specific suppression via stripping cognate peptide-MHCII from the DC surface. *J. Immunol.* **200**, 47.20-47.20 (2018).

238. Akkaya, B. *et al.* Regulatory T cells mediate specific suppression by depleting peptide–MHC class II from dendritic cells. *Nat. Immunol.* **20**, 218–231 (2019).

239. Fallarino, F. *et al.* Modulation of tryptophan catabolism by regulatory T cells. *Nat. Immunol.* **4**, 1206–1212 (2003).

240. Mellor, A. L. & Munn, D. H. Ido expression by dendritic cells: tolerance and tryptophan catabolism. *Nat. Rev. Immunol.* **4**, 762–774 (2004).

241. Wing, K. *et al.* CTLA-4 Control over Foxp3+ Regulatory T Cell Function. *Science* **322**, 271–275 (2008).

242. Joller, N. & Kuchroo, V. K. Tim-3, Lag-3, and TIGIT. *Curr. Top. Microbiol. Immunol.* **410**, 127–156 (2017).

243. Gianchecchi, E. & Fierabracci, A. Inhibitory Receptors and Pathways of Lymphocytes: The Role of PD-1 in Treg Development and Their Involvement in Autoimmunity Onset and Cancer Progression. *Front. Immunol.* **9**, (2018).

244. Lieberman, J. The ABCs of granule-mediated cytotoxicity: new weapons in the arsenal. *Nat. Rev. Immunol.* **3**, 361–370 (2003).

245. Grossman, W. J. *et al.* Human T Regulatory Cells Can Use the Perforin Pathway to Cause Autologous Target Cell Death. *Immunity* **21**, 589–601 (2004).

246. Cao, X. *et al.* Granzyme B and perforin are important for regulatory T cell-mediated suppression of tumor clearance. *Immunity* **27**, 635–646 (2007).

247. Ren, X. *et al.* Involvement of cellular death in TRAIL/DR5-dependent suppression induced by CD4 + CD25 + regulatory T cells. *Cell Death Differ.* **14**, 2076–2084 (2007).

248. Bodmer, J.-L. *et al.* TRAIL receptor-2 signals apoptosis through FADD and caspase-8. *Nat. Cell Biol.* **2**, 241–243 (2000).

249. Groux, H. *et al.* A CD4 + T-cell subset inhibits antigen-specific T-cell responses and prevents colitis. *Nature* **389**, 737–742 (1997).

250. Gagliani, N. *et al.* Coexpression of CD49b and LAG-3 identifies human and mouse T regulatory type 1 cells. *Nat. Med.* **19**, 739–746 (2013).

251. Barrat, F. J. *et al.* In Vitro Generation of Interleukin 10–producing Regulatory CD4+ T Cells Is Induced by Immunosuppressive Drugs and Inhibited by T Helper Type 1 (Th1)– and Th2-inducing Cytokines. *J. Exp. Med.* **195**, 603–616 (2002).

252. Awasthi, A. *et al.* A dominant function for interleukin 27 in generating interleukin 10–producing anti-inflammatory T cells. *Nat. Immunol.* **8**, 1380–1389 (2007).

253. Fitzgerald, D. C. *et al.* Suppression of autoimmune inflammation of the central nervous system by interleukin 10 secreted by interleukin 27–stimulated T cells. *Nat. Immunol.* **8**, 1372–1379 (2007).

254. Stumhofer, J. S. *et al.* Interleukins 27 and 6 induce STAT3-mediated T cell production of interleukin 10. *Nat. Immunol.* **8**, 1363–1371 (2007).

255. Yoshida, H. & Hunter, C. A. The Immunobiology of Interleukin-27. *Annu. Rev. Immunol.* **33**, 417–443 (2015).

256. Iwasaki, Y. *et al.* Egr-2 transcription factor is required for Blimp-1-mediated IL-10 production in IL-27-stimulated CD4+ T cells. *Eur. J. Immunol.* **43**, 1063–1073 (2013).

257. Pot, C. *et al.* Cutting Edge: IL-27 Induces the Transcription Factor c-Maf, Cytokine IL-21, and the Costimulatory Receptor ICOS that Coordinately Act Together to Promote Differentiation of IL-10-Producing Tr1 Cells. *J. Immunol.* **183**, 797–801 (2009).

258. Apetoh, L. *et al.* The aryl hydrocarbon receptor interacts with c-Maf to promote the differentiation of type 1 regulatory T cells induced by IL-27. *Nat. Immunol.* **11**, 854–861 (2010).

259. Weiner, H. L., Cunha, A. P. da, Quintana, F. & Wu, H. Oral tolerance. *Immunol. Rev.* **241**, 241–259 (2011).

260. Weiner, H. L. Induction and mechanism of action of transforming growth factor-β-secreting Th3 regulatory cells. *Immunol. Rev.* **182**, 207–214 (2001).

261. Elkord, E., Samid, M. A. A. & Chaudhary, B. Helios, and not FoxP3, is the marker of activated Tregs expressing GARP/LAP. *Oncotarget* **6**, 20026–20036 (2015).

262. Inobe, J. *et al.* IL-4 is a differentiation factor for transforming growth factor-β secreting Th3 cells and oral administration of IL-4 enhances oral tolerance in experimental allergic encephalomyelitis. *Eur. J. Immunol.* **28**, 2780–2790 (1998).

263. Chaudhry, A. *et al.* CD4+ Regulatory T Cells Control TH17 Responses in a Stat3-Dependent Manner. *Science* **326**, 986–991 (2009).

264. Chung, Y. *et al.* Follicular regulatory T cells expressing Foxp3 and Bcl-6 suppress germinal center reactions. *Nat. Med.* **17**, 983–988 (2011).
265. Koch, M. A. *et al.* The transcription factor T-bet controls regulatory T cell homeostasis and function during type 1 inflammation. *Nat. Immunol.* **10**, 595–602 (2009).
266. Linterman, M. A. *et al.* Foxp3+ follicular regulatory T cells control the germinal center response. *Nat. Med. N. Y.* **17**, 975–82 (2011).
267. Wohlfert, E. A. *et al.* GATA3 controls Foxp3+ regulatory T cell fate during inflammation in mice. *J. Clin. Invest.* **121**, 4503–4515 (2011).
268. Ohnmacht, C. *et al.* The microbiota regulates type 2 immunity through RORγt+ T cells. *Science* **349**, 989–993 (2015).
269. Sefik, E. *et al.* Individual intestinal symbionts induce a distinct population of RORγ+ regulatory T cells. *Science* **349**, 993–997 (2015).
270. Zheng, Y. *et al.* Regulatory T-cell suppressor program co-opts transcription factor IRF4 to control TH2 responses. *Nature* **458**, 351–356 (2009).
271. Hall, A. O. *et al.* The cytokines Interleukin 27 and Interferon-γ promote distinct Treg cell populations required to limit infection-induced pathology. *Immunity* **37**, 511–523 (2012).
272. Bettelli, E. *et al.* Loss of T-bet, But Not STAT1, Prevents the Development of Experimental Autoimmune Encephalomyelitis. *J. Exp. Med.* **200**, 79–87 (2004).
273. Levine, A. G. *et al.* Stability and function of regulatory T cells expressing the transcription factor T-bet. *Nature* (2017) doi:10.1038/nature22360.
274. Paust, H.-J. *et al.* CXCR3+ Regulatory T Cells Control TH1 Responses in Crescentic GN. *J. Am. Soc. Nephrol.* **27**, 1933–1942 (2016).
275. Jin, H., Park, Y., Elly, C. & Liu, Y.-C. Itch expression by Treg cells controls Th2 inflammatory responses. *J. Clin. Invest.* **123**, 4923–4934 (2013).
276. MacDonald, K. G. *et al.* Regulatory T cells produce profibrotic cytokines in the skin of patients with systemic sclerosis. *J. Allergy Clin. Immunol.* **135**, 946-955.e9 (2015).
277. Rivas, M. N. *et al.* Regulatory T cell reprogramming towards a Th2 cell-like lineage impairs oral tolerance and promotes food allergy. *Immunity* **42**, 512–523 (2015).
278. Moosbrugger-Martinz, V., Tripp, C. H., Clausen, B. E., Schmuth, M. & Dubrac, S. Atopic dermatitis induces the expansion of thymus-derived regulatory T cells exhibiting a Th2-like phenotype in mice. *J. Cell. Mol. Med.* **20**, 930–938 (2016).
279. Lohoff, M. *et al.* Dysregulated T helper cell differentiation in the absence of interferon regulatory factor 4. *Proc. Natl. Acad. Sci.* **99**, 11808–11812 (2002).
280. Cretney, E. *et al.* The transcription factors Blimp-1 and IRF4 jointly control the differentiation and function of effector regulatory T cells. *Nat. Immunol.* **12**, 304–311 (2011).
281. Koizumi, S. *et al.* JunB regulates homeostasis and suppressive functions of effector regulatory T cells. *Nat. Commun.* **9**, 1–14 (2018).
282. Chen, Q. *et al.* ICOS signal facilitates Foxp3 transcription to favor suppressive function of regulatory T cells. *Int. J. Med. Sci.* **15**, 666–673 (2018).
283. Campbell, D. J. Regulatory T Cells GATA Have It. *Immunity* **35**, 313–315 (2011).
284. Wang, Y., Su, M. A. & Wan, Y. Y. An Essential Role of the Transcription Factor GATA-3 for the Function of Regulatory T Cells. *Immunity* **35**, 337–348 (2011).
285. Wei, G. *et al.* Genome-wide Analyses of Transcription Factor GATA3-Mediated Gene Regulation in Distinct T Cell Types. *Immunity* **35**, 299–311 (2011).
286. Wan, Y. Y. & Flavell, R. A. Regulatory T-cell functions are subverted and converted owing to attenuated Foxp3 expression. *Nature* **445**, 766–770 (2007).
287. Yu, F., Sharma, S., Edwards, J., Feigenbaum, L. & Zhu, J. Dynamic expression of transcription factors T-bet and GATA-3 by regulatory T cells maintains immunotolerance. *Nat. Immunol.* **16**, 197–206 (2015).
288. Zhou, L. *et al.* TGF-β-induced Foxp3 inhibits Th17 cell differentiation by antagonizing RORγt function. *Nature* **453**, 236–240 (2008).
289. Lochner, M. *et al.* In vivo equilibrium of proinflammatory IL-17+ and regulatory IL-10+ Foxp3+ RORγt+ T cells. *J. Exp. Med.* **205**, 1381–1393 (2008).
290. Osorio, F. *et al.* DC activated via dectin-1 convert Treg into IL-17 producers. *Eur. J. Immunol.* **38**, 3274–3281 (2008).
291. Ayyoub, M. *et al.* Human memory FOXP3+ Tregs secrete IL-17 ex vivo and constitutively express the TH17 lineage-specific transcription factor RORγt. *Proc. Natl. Acad. Sci.* **106**, 8635–8640 (2009).
292. Beriou, G. *et al.* IL-17–producing human peripheral regulatory T cells retain suppressive function. *Blood* **113**, 4240–4249 (2009).
293. Gounaris, E. *et al.* T-Regulatory Cells Shift from a Protective Anti-Inflammatory to a Cancer-Promoting Proinflammatory Phenotype in Polyposis. *Cancer Res.* **69**, 5490–5497 (2009).
294. Voo, K. S. *et al.* Identification of IL-17-producing FOXP3+ regulatory T cells in humans. *Proc. Natl. Acad. Sci. U. S. A.* **106**, 4793–4798 (2009).

295. Tartar, D. M. *et al.* FoxP3+RORγt+ T helper intermediates display suppressive function against autoimmune diabetes. *J. Immunol. Baltim. Md 1950* **184**, 3377–3385 (2010).

296. Kryczek, I. *et al.* IL-17+ Regulatory T Cells in the Microenvironments of Chronic Inflammation and Cancer. *J. Immunol.* **186**, 4388–4395 (2011).

297. Blatner, N. R. *et al.* Expression of RORγt Marks a Pathogenic Regulatory T Cell Subset in Human Colon Cancer. *Sci. Transl. Med.* **4**, 164ra159-164ra159 (2012).

298. Chellappa, S. *et al.* Regulatory T cells that co-express RORγt and FOXP3 are pro-inflammatory and immunosuppressive and expand in human pancreatic cancer. *Oncoimmunology* **5**, (2015).

299. Phillips, J. D. *et al.* Preferential expansion of pro-inflammatory Tregs in human non-small cell lung cancer. *Cancer Immunol. Immunother. CII* **64**, 1185–1191 (2015).

300. Kluger, M. A. *et al.* RORγt+Foxp3+ Cells are an Independent Bifunctional Regulatory T Cell Lineage and Mediate Crescentic GN. *J. Am. Soc. Nephrol.* **27**, 454–465 (2016).

301. Solomon, B. D. & Hsieh, C.-S. Antigen-Specific Development of Mucosal Foxp3+RORγt+ T Cells from Regulatory T Cell Precursors. *J. Immunol.* 1601217 (2016) doi:10.4049/jimmunol.1601217.

302. Timperi, E. *et al.* Regulatory T cells with multiple suppressive and potentially pro-tumor activities accumulate in human colorectal cancer. *Oncoimmunology* **5**, (2016).

303. Yang, B.-H. *et al.* Foxp3+ T cells expressing RORγt represent a stable regulatory T-cell effector lineage with enhanced suppressive capacity during intestinal inflammation. *Mucosal Immunol.* **9**, 444–457 (2016).

304. Kim, B.-S. *et al.* Generation of RORγt+ antigen-specific T regulatory 17 (Tr17) cells from Foxp3+ precursors in autoimmunity. *Cell Rep.* **21**, 195–207 (2017).

305. Wheaton, J. D., Yeh, C.-H. & Ciofani, M. Cutting Edge: c-Maf Is Required for Regulatory T Cells To Adopt RORγt+ and Follicular Phenotypes. *J. Immunol.* ji1701134 (2017) doi:10.4049/jimmunol.1701134.

306. Tosiek, M. J., Fiette, L., El Daker, S., Eberl, G. & Freitas, A. A. IL-15-dependent balance between Foxp3 and RORγt expression impacts inflammatory bowel disease. *Nat. Commun.* **7**, (2016).

307. Xu, M. *et al.* c-MAF-dependent regulatory T cells mediate immunological tolerance to a gut pathobiont. *Nature* **554**, 373–377 (2018).

308. Imbratta, C. *et al.* Maf deficiency in T cells dysregulates T reg - T H 17 balance leading to spontaneous colitis. *Sci. Rep.* **9**, 6135 (2019).

309. Neumann, C. *et al.* c-Maf-dependent T reg cell control of intestinal T H 17 cells and IgA establishes host–microbiota homeostasis. *Nat. Immunol.* 1 (2019) doi:10.1038/s41590-019-0316-2.

310. Hussein, H. *et al.* Multiple Environmental Signaling Pathways Control the Differentiation of RORγt-Expressing Regulatory T Cells. *Front. Immunol.* **10**, (2020).

311. Hirota, K. *et al.* Preferential recruitment of CCR6-expressing Th17 cells to inflamed joints via CCL20 in rheumatoid arthritis and its animal model. *J. Exp. Med.* **204**, 2803–2812 (2007).

312. Yamazaki, T. *et al.* CCR6 Regulates the Migration of Inflammatory and Regulatory T Cells. *J. Immunol.* **181**, 8391–8401 (2008).

313. Kitamura, K., Farber, J. M. & Kelsall, B. L. CCR6 marks regulatory T cells as a colon-tropic, interleukin-10-producing phenotype. *J. Immunol. Baltim. Md 1950* **185**, 3295–3304 (2010).

314. Villares, R. *et al.* CCR6 regulates EAE pathogenesis by controlling regulatory CD4+ T-cell recruitment to target tissues. *Eur. J. Immunol.* **39**, 1671–1681 (2009).

315. Wollenberg, I. *et al.* Regulation of the Germinal Center Reaction by Foxp3+ Follicular Regulatory T Cells. *J. Immunol.* **187**, 4553–4560 (2011).

316. Laidlaw, B. J. *et al.* Interleukin-10 from CD4+ follicular regulatory T cells promotes the germinal center response. *Sci. Immunol.* **2**, (2017).

317. Xiong, N. & Hu, S. Regulation of intestinal IgA responses. *Cell. Mol. Life Sci.* **72**, 2645–2655 (2015).

318. Wu, H., Xie, M. M., Liu, H. & Dent, A. L. Stat3 Is Important for Follicular Regulatory T Cell Differentiation. *PLoS ONE* **11**, (2016).

319. Dominguez-Villar, M., Baecher-Allan, C. M. & A Hafler, D. Identification of T helper type 1–like, Foxp3+ regulatory T cells in human autoimmune disease. *Nat. Med.* **17**, 673–675 (2011).

320. Pasare, C. & Medzhitov, R. Toll Pathway-Dependent Blockade of CD4+CD25+ T Cell-Mediated Suppression by Dendritic Cells. *Science* **299**, 1033–1036 (2003).

321. Yang, X. O. *et al.* Molecular antagonism and plasticity of regulatory and inflammatory T cell programs. *Immunity* **29**, 44–56 (2008).

322. Korn, T. *et al.* Myelin-specific regulatory T cells accumulate in the CNS but fail to control autoimmune inflammation. *Nat. Med.* **13**, 423–431 (2007).

323. Oldenhove, G. *et al.* Decrease of Foxp3+ Treg cell number and acquisition of effector cell phenotype during lethal infection. *Immunity* **31**, 772 (2009).

324. Feuerer, M. *et al.* Lean, but not obese, fat is enriched for a unique population of regulatory T cells that affect metabolic parameters. *Nat. Med.* **15**, 930–939 (2009).

325. Cipolletta, D. *et al.* PPAR-γ is a major driver of the accumulation and phenotype of adipose tissue T reg cells. *Nature* **486**, 549–553 (2012).

326. Vasanthakumar, A. *et al.* The transcriptional regulators IRF4, BATF and IL-33 orchestrate development and maintenance of adipose tissue–resident regulatory T cells. *Nat. Immunol.* **16**, 276–285 (2015).

327. Ali, N. *et al.* Regulatory T Cells in Skin Facilitate Epithelial Stem Cell Differentiation. *Cell* **169**, 1119-1129.e11 (2017).

328. Burzyn, D. *et al.* A Special Population of Regulatory T Cells Potentiates Muscle Repair. *Cell* **155**, 1282–1295 (2013).

329. Arpaia, N. *et al.* A Distinct Function of Regulatory T Cells in Tissue Protection. *Cell* **162**, 1078–1089 (2015).

330. Zaiss, D. M. W., Gause, W. C., Osborne, L. C. & Artis, D. Emerging functions of amphiregulin in orchestrating immunity, inflammation and tissue repair. *Immunity* **42**, 216–226 (2015).

331. Schiering, C. *et al.* The Alarmin IL-33 Promotes Regulatory T Cell Function in the Intestine. *Nature* **513**, 564–568 (2014).

332. Harrison, O. J. & Powrie, F. M. Regulatory T Cells and Immune Tolerance in the Intestine. *Cold Spring Harb. Perspect. Biol.* **5**, (2013).

333. Mestecky, J. & Elson, C. O. Peyer's Patches as the Inductive Site for IgA Responses. *J. Immunol.* **180**, 1293–1294 (2008).

334. Sharma, A. & Rudra, D. Emerging Functions of Regulatory T Cells in Tissue Homeostasis. *Front. Immunol.* **9**, (2018).

335. Bevins, C. L. & Salzman, N. H. Paneth cells, antimicrobial peptides and maintenance of intestinal homeostasis. *Nat. Rev. Microbiol.* **9**, 356–368 (2011).

336. Pelaseyed, T. *et al.* The mucus and mucins of the goblet cells and enterocytes provide the first defense line of the gastrointestinal tract and interact with the immune system. *Immunol. Rev.* **260**, 8–20 (2014).

337. Gribble, F. M. & Reimann, F. Function and mechanisms of enteroendocrine cells and gut hormones in metabolism. *Nat. Rev. Endocrinol.* **15**, 226–237 (2019).

338. Schneider, C., O'Leary, C. E. & Locksley, R. M. Regulation of immune responses by tuft cells. *Nat. Rev. Immunol.* **19**, 584–593 (2019).

339. Olivares-Villagómez, D. & Kaer, L. V. Intestinal Intraepithelial Lymphocytes: Sentinels of the Mucosal Barrier. *Trends Immunol.* **39**, 264–275 (2018).

340. Sender, R., Fuchs, S. & Milo, R. Revised Estimates for the Number of Human and Bacteria Cells in the Body. *PLoS Biol.* **14**, (2016).

341. Blacher, E., Levy, M., Tatirovsky, E. & Elinav, E. Microbiome-Modulated Metabolites at the Interface of Host Immunity. *J. Immunol.* **198**, 572–580 (2017).

342. McCarville, J. L., Chen, G. Y., Cuevas, V. D., Troha, K. & Ayres, J. S. Microbiota Metabolites in Health and Disease. *Annu. Rev. Immunol.* **38**, 147–170 (2020).

343. Kayama, H., Okumura, R. & Takeda, K. Interaction Between the Microbiota, Epithelia, and Immune Cells in the Intestine. *Annu. Rev. Immunol.* **38**, 23–48 (2020).

344. Chung, H. *et al.* Gut Immune Maturation Depends on Colonization with a Host-Specific Microbiota. *Cell* **149**, 1578–1593 (2012).

345. Annunziato, F., Romagnani, C. & Romagnani, S. The 3 major types of innate and adaptive cell-mediated effector immunity. *J. Allergy Clin. Immunol.* **135**, 626–635 (2015).

346. Ohnmacht, C. Tolerance to the Intestinal Microbiota Mediated by ROR(γt)+ Cells. *Trends Immunol.* **37**, 477–486 (2016).

347. Cassani, B. *et al.* Gut-tropic T Cells that Express Integrin α4β7 and CCR9 are Required for Induction of Oral Immune Tolerance in Mice. *Gastroenterology* **141**, 2109–2118 (2011).

348. Omenetti, S. *et al.* The Intestine Harbors Functionally Distinct Homeostatic Tissue-Resident and Inflammatory Th17 Cells. *Immunity* **51**, 77-89.e6 (2019).

349. Round, J. L. & Mazmanian, S. K. The gut microbiota shapes intestinal immune responses during health and disease. *Nat. Rev. Immunol.* **9**, 313–323 (2009).

350. Stagg, A. J. Intestinal Dendritic Cells in Health and Gut Inflammation. *Front. Immunol.* **9**, (2018).

351. Kim, K. S. *et al.* Dietary antigens limit mucosal immunity by inducing regulatory T cells in the small intestine. *Science* **351**, 858–863 (2016).

352. Siede, J. *et al.* IL-33 Receptor-Expressing Regulatory T Cells Are Highly Activated, Th2 Biased and Suppress CD4 T Cell Proliferation through IL-10 and TGFβ Release. *PLOS ONE* **11**, e0161507 (2016).

353. Bertheloot, D. & Latz, E. HMGB1, IL-1α, IL-33 and S100 proteins: dual-function alarmins. *Cell. Mol. Immunol.* **14**, 43–64 (2017).

354. Chan, J. K. *et al.* Alarmins: awaiting a clinical response. *J. Clin. Invest.* **122**, 2711–2719 (2012).

355. Zaiss, D. M. W. *et al.* Amphiregulin Enhances Regulatory T Cell-Suppressive Function via the Epidermal Growth Factor Receptor. *Immunity* **38**, 275–284 (2013).

356. Atarashi, K. *et al.* Induction of Colonic Regulatory T Cells by Indigenous Clostridium Species. *Science* **331**, 337–341 (2011).

357. Geuking, M. B. *et al.* Intestinal Bacterial Colonization Induces Mutualistic Regulatory T Cell Responses. *Immunity* **34**, 794–806 (2011).

358. Frank, D. N. *et al.* Molecular-phylogenetic characterization of microbial community imbalances in human inflammatory bowel diseases. *Proc. Natl. Acad. Sci.* **104**, 13780–13785 (2007).

359. R, T. *et al.* Down-regulation of the monocarboxylate transporter 1 is involved in butyrate deficiency during intestinal inflammation. *Gastroenterology* **133**, 1916–1927 (2007).

360. Enzinger, F. M. & Shiraki, M. Musculo-aponeurotic fibromatosis of the shoulder girdle (extra-abdominal desmoid). Analysis of thirty cases followed up for ten or more years. *Cancer* **20**, 1131–1140 (1967).

361. Nishizawa, M., Kataoka, K., Goto, N., Fujiwara, K. T. & Kawai, S. v-maf, a viral oncogene that encodes a 'leucine zipper' motif. *Proc. Natl. Acad. Sci. U. S. A.* **86**, 7711–7715 (1989).

362. Kawai, S. *et al.* Isolation of the avian transforming retrovirus, AS42, carrying the v-maf oncogene and initial characterization of its gene product. *Virology* **188**, 778–784 (1992).

363. Kawauchi, S. *et al.* Regulation of lens fiber cell differentiation by transcription factor c-Maf. *J. Biol. Chem.* **274**, 19254–19260 (1999).

364. Ring, B. Z., Cordes, S. P., Overbeek, P. A. & Barsh, G. S. Regulation of mouse lens fiber cell development and differentiation by the Maf gene. *Development* **127**, 307–317 (2000).

365. Kim, J. I., Li, T., Ho, I.-C., Grusby, M. J. & Glimcher, L. H. Requirement for the c-Maf transcription factor in crystallin gene regulation and lens development. *Proc. Natl. Acad. Sci.* **96**, 3781–3785 (1999).

366. Wende, H. *et al.* The Transcription Factor c-Maf Controls Touch Receptor Development and Function. *Science* **335**, 1373–1376 (2012).

367. Imaki, J. *et al.* Developmental contribution of c-maf in the kidney: distribution and developmental study of c-maf mRNA in normal mice kidney and histological study of c-maf knockout mice kidney and liver. *Biochem. Biophys. Res. Commun.* **320**, 1323–1327 (2004).

368. Huang, W., Lu, N., Eberspaecher, H. & Crombrugghe, B. de. A New Long Form of c-Maf Cooperates with Sox9 to Activate the Type II Collagen Gene. *J. Biol. Chem.* **277**, 50668–50675 (2002).

369. Hong, E., Di Cesare, P. E. & Haudenschild, D. R. Role of c-Maf in Chondrocyte Differentiation. *Cartilage* **2**, 27–35 (2011).

370. MacLean, H. E. *et al.* Absence of transcription factor c-maf causes abnormal terminal differentiation of hypertrophic chondrocytes during endochondral bone development. *Dev. Biol.* **262**, 51–63 (2003).

371. Kusakabe, M. *et al.* c-Maf plays a crucial role for the definitive erythropoiesis that accompanies erythroblastic island formation in the fetal liver. *Blood* **118**, 1374–1385 (2011).

372. Kataoka, K. *et al.* Differentially expressed Maf family transcription factors, c-Maf and MafA, activate glucagon and insulin gene expression in pancreatic islet alpha- and beta-cells. *J. Mol. Endocrinol.* **32**, 9–20 (2004).

373. Tsuchiya, M. *et al.* Potential Roles of Large Mafs in Cell Lineages and Developing Pancreas. *Pancreas* **32**, 408–416 (2006).

374. Wende, H., Lechner, S. G. & Birchmeier, C. The transcription factor c-Maf in sensory neuron development. *Transcription* **3**, 285–289 (2012).

375. Vinson, C., Acharya, A. & Taparowsky, E. J. Deciphering B-ZIP transcription factor interactions in vitro and in vivo. *Biochim. Biophys. Acta BBA - Gene Struct. Expr.* **1759**, 4–12 (2006).

376. Kerppola, T. K. & Curran, T. Maf and Nrl can bind to AP-1 sites and form heterodimers with Fos and Jun. *Oncogene* **9**, 675–684 (1994).

377. Kataoka, K., Fujiwara, K. T., Noda, M. & Nishizawa, M. MafB, a new Maf family transcription activator that can associate with Maf and Fos but not with Jun. *Mol. Cell. Biol.* **14**, 7581–7591 (1994).

378. Kataoka, K. Multiple Mechanisms and Functions of Maf Transcription Factors in the Regulation of Tissue-Specific Genes. *J. Biochem. (Tokyo)* **141**, 775–781 (2007).

379. Yang, Y., Ochando, J., Yopp, A., Bromberg, J. S. & Ding, Y. IL-6 Plays a Unique Role in Initiating c-Maf Expression during Early Stage of CD4 T Cell Activation. *J. Immunol.* **174**, 2720–2729 (2005).

380. Xu, J. *et al.* c-Maf Regulates IL-10 Expression during Th17 Polarization. *J. Immunol. Baltim. Md 1950* **182**, 6226–6236 (2009).

381. Rutz, S. *et al.* Transcription factor c-Maf mediates the TGF-β-dependent suppression of IL-22 production in TH17 cells. *Nat. Immunol.* **12**, 1238–1245 (2011).

382. Kurata, H., Lee, H. J., O'Garra, A. & Arai, N. Ectopic Expression of Activated Stat6 Induces the Expression of Th2-Specific Cytokines and Transcription Factors in Developing Th1 Cells. *Immunity* **11**, 677–688 (1999).

383. Ouyang, W. *et al.* Stat6-Independent GATA-3 Autoactivation Directs IL-4-Independent Th2 Development and Commitment. *Immunity* **12**, 27–37 (2000).

384. Coyle, A. J. *et al.* The CD28-Related Molecule ICOS Is Required for Effective T Cell–Dependent Immune Responses. *Immunity* **13**, 95–105 (2000).

385. Nurieva, R. I. *et al.* Transcriptional Regulation of Th2 Differentiation by Inducible Costimulator. *Immunity* **18**, 801–811 (2003).

386. Paulos, C. M. *et al.* The Inducible Costimulator (ICOS) Is Critical for the Development of Human TH17 Cells. *Sci. Transl. Med.* **2**, 55ra78 (2010).

387. Rodriguez, A. *et al.* Requirement of bic/microRNA-155 for Normal Immune Function. *Science* **316**, 608–611 (2007).

388. Su, W. *et al.* The p53 transcription factor modulates microglia behavior through microRNA dependent regulation of c-Maf. *J. Immunol. Baltim. Md 1950* **192**, 358–366 (2014).

389. Tamgue, O. *et al.* Differential Targeting of c-Maf, Bach-1, and Elmo-1 by microRNA-143 and microRNA-365 Promotes the Intracellular Growth of Mycobacterium tuberculosis in Alternatively IL-4/IL-13 Activated Macrophages. *Front. Immunol.* **10**, (2019).

390. Janiszewska, J. *et al.* Global miRNA Expression Profiling Identifies miR-1290 as Novel Potential oncomiR in Laryngeal Carcinoma. *PLoS ONE* **10**, (2015).

391. Blonska, M. *et al.* Activation of the Transcription Factor c-Maf in T Cells Is Dependent on the CARMA1-IKKβ Signaling Cascade. *Sci. Signal.* **6**, ra110 (2013).

392. Qiang, Y.-W. *et al.* MAF protein mediates innate resistance to proteasome inhibition therapy in multiple myeloma. *Blood* **128**, 2919–2930 (2016).

393. Niceta, M. *et al.* Mutations Impairing GSK3-Mediated MAF Phosphorylation Cause Cataract, Deafness, Intellectual Disability, Seizures, and a Down Syndrome-like Facies. *Am. J. Hum. Genet.* **96**, 816–825 (2015).

394. Hill, E. V. *et al.* Glycogen synthase kinase-3 controls IL-10 expression in CD4+ effector T-cell subsets through epigenetic modification of the IL-10 promoter. *Eur. J. Immunol.* **45**, 1103–1115 (2015).

395. Liu, C.-C. *et al.* Reciprocal Regulation of C-Maf Tyrosine Phosphorylation by Tec and Ptpn22. *PLoS ONE* **10**, (2015).

396. Leavenworth, J. W., Ma, X., Mo, Y. & Pauza, M. E. SUMO Conjugation Contributes to Immune Deviation in Nonobese Diabetic Mice by Suppressing c-Maf Transactivation of IL-4. *J. Immunol. Baltim. Md 1950* **183**, 1110–1119 (2009).

397. Lin, B.-S. *et al.* SUMOylation attenuates c-Maf-dependent IL-4 expression. *Eur. J. Immunol.* **40**, 1174–1184 (2010).

398. Ho, I.-C., Hodge, M. R., Rooney, J. W. & Glimcher, L. H. The Proto-Oncogene c-maf Is Responsible for Tissue-Specific Expression of Interleukin-4. *Cell* **85**, 973–983 (1996).

399. Hodge, M. R. *et al.* NF-AT-Driven Interleukin-4 Transcription Potentiated by NIP45. *Science* **274**, 1903–1905 (1996).

400. Li, B., Tournier, C., Davis, R. J. & Flavell, R. A. Regulation of IL-4 expression by the transcription factor JunB during T helper cell differentiation. *EMBO J.* **18**, 420–432 (1999).

401. Gabryšová, L. *et al.* c-Maf controls immune responses by regulating disease-specific gene networks and repressing IL-2 in CD4 + T cells. *Nat. Immunol.* **19**, 497–507 (2018).

402. Ho, I.-C., Lo, D. & Glimcher, L. H. c-maf Promotes T Helper Cell Type 2 (Th2) and Attenuates Th1 Differentiation by Both Interleukin 4–dependent and –independent Mechanisms. *J. Exp. Med.* **188**, 1859–1866 (1998).

403. Mitchell, R. E. *et al.* IL-4 enhances IL-10 production in Th1 cells: implications for Th1 and Th2 regulation. *Sci. Rep.* **7**, 1–14 (2017).

404. Tanaka, S. *et al.* Sox5 and c-Maf cooperatively induce Th17 cell differentiation via RORγt induction as downstream targets of Stat3. *J. Exp. Med.* **211**, 1857–1874 (2014).

405. Ciofani, M. *et al.* A Validated Regulatory Network for Th17 Cell Specification. *Cell* **151**, 289–303 (2012).

406. Andris, F. *et al.* The transcription factor c-Maf promotes the differentiation of follicular helper T cells. *Front. Immunol.* **8**, (2017).

407. Hiramatsu, Y. *et al.* c-Maf activates the promoter and enhancer of the IL-21 gene, and TGF-β inhibits c-Maf-induced IL-21 production in CD4+ T cells. *J. Leukoc. Biol.* **87**, 703–712 (2010).

408. Sahoo, A. *et al.* Batf is important for IL-4 expression in T follicular helper cells. *Nat. Commun.* **6**, 7997 (2015).

409. Neumann, C. *et al.* Role of Blimp-1 in programing Th effector cells into IL-10 producers. *J. Exp. Med.* **211**, 1807–1819 (2014).

410. Giordano, M. *et al.* Molecular profiling of CD8 T cells in autochthonous melanoma identifies Maf as driver of exhaustion. *EMBO J.* **34**, 2042–2058 (2015).

411. Speiser, D. E., Ho, P.-C. & Verdeil, G. Regulatory circuits of T cell function in cancer. *Nat. Rev. Immunol.* **16**, 599–611 (2016).

412. Wherry, E. J. T cell exhaustion. *Nat. Immunol.* **12**, 492–499 (2011).

413. Chihara, N. *et al.* Induction and transcriptional regulation of the co-inhibitory gene module in T cells. *Nature* **558**, 454 (2018).

414. Liston, A. & Gray, D. H. D. Homeostatic control of regulatory T cell diversity. *Nat. Rev. Immunol.* **14**, 154–165 (2014).

415. Herman, A. E., Freeman, G. J., Mathis, D. & Benoist, C. CD4+CD25+ T Regulatory Cells Dependent on ICOS Promote Regulation of Effector Cells in the Prediabetic Lesion. *J. Exp. Med.* **199**, 1479–1489 (2004).

416. Ito, T. *et al.* Two Functional Subsets of FOXP3+ Regulatory T Cells in Human Thymus and Periphery. *Immunity* **28**, 870–880 (2008).

417. Smigiel, K. S. *et al.* CCR7 provides localized access to IL-2 and defines homeostatically distinct regulatory T cell subsets. *J. Exp. Med.* **211**, 121–136 (2014).

418. Franckaert, D. *et al.* Promiscuous Foxp3-cre activity reveals a differential requirement for CD28 in Foxp3+ and Foxp3− T cells. *Immunol. Cell Biol.* **93**, 417–423 (2015).

419. Becher, B., Waisman, A. & Lu, L.-F. Conditional Gene-Targeting in Mice: Problems and Solutions. *Immunity* **48**, 835–836 (2018).

420. Wu, D., Huang, Q., Orban, P. C. & Levings, M. K. Ectopic germline recombination activity of the widely used Foxp3-YFP-Cre mouse: a case report. *Immunology* **159**, 231–241 (2020).

421. Miyara, M. *et al.* Human FoxP3+ regulatory T cells in systemic autoimmune diseases. *Autoimmun. Rev.* **10**, 744–755 (2011).
422. Ahn, J., Son, S., Oliveira, S. C. & Barber, G. N. STING-Dependent Signaling Instigates IL-10 Controlled Inflammatory Colitis. *Cell Rep.* **21**, 3873–3884 (2017).
423. Miller, C. L., Muthupalani, S., Shen, Z. & Fox, J. G. Isolation of Helicobacter spp. from Mice with Rectal Prolapses. *Comp. Med.* **64**, 171–178 (2014).
424. Wohlfert, E. A., Warunek, J., Jin, R. M. & Marzullo, B. Tbet-expressing Tregs protect against lethal immunopathology during T. gondii infection. *J. Immunol.* **204**, 232.4-232.4 (2020).
425. Sakaguchi, S., Miyara, M., Costantino, C. M. & Hafler, D. A. FOXP3+ regulatory T cells in the human immune system. *Nat. Rev. Immunol.* **10**, 490–500 (2010).
426. Togashi, Y. & Nishikawa, H. Regulatory T Cells: Molecular and Cellular Basis for Immunoregulation. in *Emerging Concepts Targeting Immune Checkpoints in Cancer and Autoimmunity* (ed. Yoshimura, A.) 3–27 (Springer International Publishing, 2017). doi:10.1007/82_2017_58.
427. Togashi, Y., Shitara, K. & Nishikawa, H. Regulatory T cells in cancer immunosuppression — implications for anticancer therapy. *Nat. Rev. Clin. Oncol.* **16**, 356–371 (2019).
428. Dahmani, A. & Delisle, J.-S. TGF-β in T Cell Biology: Implications for Cancer Immunotherapy. *Cancers* **10**, (2018).
429. Stidham, R. W. & Higgins, P. D. R. Colorectal Cancer in Inflammatory Bowel Disease. *Clin. Colon Rectal Surg.* **31**, 168–178 (2018).
430. Song, X. *et al.* Microbial bile acid metabolites modulate gut RORγ + regulatory T cell homeostasis. *Nature* **577**, 410–415 (2020).
431. Gao, Z. *et al.* Synergy between IL-6 and TGF-β signaling promotes FOXP3 degradation. *Int. J. Clin. Exp. Pathol.* **5**, 626–633 (2012).
432. Fujimoto, M. *et al.* The influence of excessive IL-6 production in vivo on the development and function of Foxp3+ regulatory T cells. *J. Immunol. Baltim. Md 1950* **186**, 32–40 (2011).
433. Hussein, H. *et al.* Multiple environmental signaling pathways control the differentiation of RORγt-expressing regulatory T cells. *Front. Immunol.* **10:3007**, (2019).
434. Panduro, M., Benoist, C. & Mathis, D. Tissue Tregs. *Annu. Rev. Immunol.* **34**, 609–633 (2016).
435. Whibley, N., Tucci, A. & Powrie, F. Regulatory T cell adaptation in the intestine and skin. *Nat. Immunol.* **20**, 386–396 (2019).
436. Macpherson, A. J., Yilmaz, B., Limenitakis, J. P. & Ganal-Vonarburg, S. C. IgA Function in Relation to the Intestinal Microbiota. *Annu. Rev. Immunol.* **36**, 359–381 (2018).
437. Wang, T. *et al.* CCR8 blockade primes anti-tumor immunity through intratumoral regulatory T cells destabilization in muscle-invasive bladder cancer. *Cancer Immunol. Immunother.* (2020) doi:10.1007/s00262-020-02583-y.
438. Parang, B., Barret, C. W. & Williams, C. S. AOM/DSS Model of Colitis-Associated Cancer. *Methods Mol. Biol. Clifton NJ* **1422**, 297–307 (2016).
439. Woude, L. L. van der, Gorris, M. A. J., Halilovic, A., Figdor, C. G. & Vries, I. J. M. de. Migrating into the Tumor: a Roadmap for T Cells. *Trends Cancer* **3**, 797–808 (2017).
440. Zhao, J., Chen, X., Herjan, T. & Li, X. The role of interleukin-17 in tumor development and progression. *J. Exp. Med.* **217**, (2020).
441. Vitiello, G. A. & Miller, G. Targeting the interleukin-17 immune axis for cancer immunotherapy. *J. Exp. Med.* **217**, (2020).
442. Chang, S. H. T helper 17 (Th17) cells and interleukin-17 (IL-17) in cancer. *Arch. Pharm. Res.* **42**, 549–559 (2019).
443. Markota, A., Endres, S. & Kobold, S. Targeting interleukin-22 for cancer therapy. *Hum. Vaccines Immunother.* **14**, 2012–2015 (2018).
444. Dmitrieva-Posocco, O. *et al.* Cell-Type-Specific Responses to Interleukin-1 Control Microbial Invasion and Tumor-Elicited Inflammation in Colorectal Cancer. *Immunity* **50**, 166-180.e7 (2019).
445. Ponzetta, A. *et al.* Neutrophils Driving Unconventional T Cells Mediate Resistance against Murine Sarcomas and Selected Human Tumors. *Cell* **178**, 346-360.e24 (2019).
446. Fong, W., Li, Q. & Yu, J. Gut microbiota modulation: a novel strategy for prevention and treatment of colorectal cancer. *Oncogene* (2020) doi:10.1038/s41388-020-1341-1.
447. Pastille, E. *et al.* Transient Ablation of Regulatory T cells Improves Antitumor Immunity in Colitis-Associated Colon Cancer. *Cancer Res.* **74**, 4258–4269 (2014).
448. Rizzo, A. *et al.* RORγt-Expressing Tregs Drive the Growth of Colitis-Associated Colorectal Cancer by Controlling IL6 in Dendritic Cells. *Cancer Immunol. Res.* **6**, 1082–1092 (2018).
449. Imbratta, C., Hussein, H., Andris, F. & Verdeil, G. c-MAF, a Swiss Army Knife for Tolerance in Lymphocytes. *Front. Immunol.* **11**, (2020).

Chapter 8: Annexes

8.1. Annex 1 – Literature review: "c-MAF, a Swiss Army Knife for Tolerance in Lymphocytes"

The role attributed to transcription factor c-Maf has gradually broadened over the years and now extends to most, if not all, known immune cell types. The influence of c-Maf is particularly prominent among T cell subsets, where c-Maf regulates the differentiation as well as the function of multiple subsets of CD4 and CD8 T cells, lending it a crucial position in adaptive immunity and anti-tumoral responsiveness.

We published a literature review, in collaboration with G. Verdeil and C. Imbratta (UNIL, Lausanne, Switzerland), describing the roles of transcription factor c-Maf among T cells. This work was published in Frontiers of Immunology on January 14th 2020 and was titled "c-MAF, a Swiss Army Knife for Tolerance in Lymphocytes"[449].

REVIEW
published: 14 February 2020
doi: 10.3389/fimmu.2020.00206

c-MAF, a Swiss Army Knife for Tolerance in Lymphocytes

Claire Imbratta[1], Hind Hussein[2], Fabienne Andris[2*†] and Grégory Verdeil[1*†]

[1] Department of Oncology, University of Lausanne, Lausanne, Switzerland, [2] Laboratoire d'Immunobiologie, Université Libre de Bruxelles, Brussels, Belgium

Beyond its well-admitted role in development and organogenesis, it is now clear that the transcription factor c-Maf has owned its place in the realm of immune-related transcription factors. Formerly introduced solely as a Th2 transcription factor, the role attributed to c-Maf has gradually broadened over the years and has extended to most, if not all, known immune cell types. The influence of c-Maf is particularly prominent among T cell subsets, where c-Maf regulates the differentiation as well as the function of multiple subsets of CD4 and CD8 T cells, lending it a crucial position in adaptive immunity and anti-tumoral responsiveness. Recent research has also revealed the role of c-Maf in controlling Th17 responses in the intestine, positioning it as an essential factor in intestinal homeostasis. This review aims to present and discuss the recent advances highlighting the particular role played by c-Maf in T lymphocyte differentiation, function, and homeostasis.

Keywords: T cells, tolerance, c-Maf, interleukin-10, gut

OPEN ACCESS

Edited by:
Remy Bosselut,
National Cancer Institute (NCI),
United States

Reviewed by:
Michele Kay Anderson,
University of Toronto, Canada
Booki Min,
Case Western Reserve University,
United States

***Correspondence:**
Fabienne Andris
fabienne.andris@ulb.be
Grégory Verdeil
gregory.verdeil@unil.ch

†These authors share last authorship

Specialty section:
This article was submitted to
T Cell Biology,
a section of the journal
Frontiers in Immunology

Received: 16 December 2019
Accepted: 27 January 2020
Published: 14 February 2020

Citation:
Imbratta C, Hussein H, Andris F and
Verdeil G (2020) c-MAF, a Swiss Army
Knife for Tolerance in Lymphocytes.
Front. Immunol. 11:206.
doi: 10.3389/fimmu.2020.00206

INTRODUCTION

The *Maf* (musculoaponeurotic fibrosarcoma) gene encodes the transcription factor c-Maf or MAF. Originally identified in natural musculo-aponeurotic fibrosarcoma of chickens infected with the replication-defective retrovirus AS42, the founding member of the Maf family, named v-Maf, was described as an oncogene (1–3). Using a probe containing the v-Maf sequence, its cellular counterpart, identified as c-Maf, was thereafter cloned from a number of vertebrate genomes (4). In addition to its function as an oncogene, c-Maf was soon found to regulate various cellular differentiation and developmental processes within tissues. In particular, c-Maf expression controls lens fiber cell differentiation, crystalline gene expression, as well as lens development (5–7). In neural tissue, c-Maf controls the expression of mechanoreceptors involved in touch sensation (8, 9). It also regulates the embryonic development of tubular renal cells (10) and the differentiation of chondrocytes during endochondral bone development (11–13). c-Maf plays a predominant role for the erythropoiesis that accompanies erythroblastic islands formation in fetal liver (14). In porcine and human pancreatic islets (15), c-Maf also regulates glucagon hormone production, thereby establishing pancreatic endocrine function (16). In line with the major contributions of c-Maf in developmental and physiological processes, mice lacking c-Maf are embryonically (14) or perinatally (5, 7) lethal depending on the type of C57BL/6 background. Some mice on the BALB/c background live to adulthood (10, 13).

In parallel to the discovery of the many roles of c-Maf within tissue development, c-Maf soon emerged as an immune regulator and was initially identified as a Th2 transcription factor. Similar to its function in tissue development, the role attributed to c-Maf within immune regulation broadened over the years and has extended to most, if not all, known immune cell types. While the role of c-Maf has also been studied within innate immune cell types (17–19) and B lymphocytes (20), we focus on c-Maf within T cell subsets, where c-Maf regulates the differentiation as well as the function of multiple subsets of CD4 T cells, lending it a crucial position in T cell immunity. Recent research has revealed the role of c-Maf in the control of intestinal Th17 responses by regulatory T cells, positioning it as an essential factor in regulatory T cell specification and, more broadly, the maintenance of intestinal homeostasis. This review aims to present and discuss the recent advances highlighting the particular role played by c-Maf in T lymphocyte differentiation, function, and homeostasis.

THE c-MAF TRANSCRIPTION FACTOR

This basic leucine zipper (bZIP) transcription factor belongs to the AP-1 superfamily, which includes Fos, Jun, ATF, and CREB. The Maf transcription factor family is composed of 7 members divided into two subclasses: the large Maf proteins composed of MAFA/L-MAF, MAFB, MAF/c-Maf, and NLR (neural retina leucine zipper), and the small Maf proteins, MAFK, MAFG, and MAFF, which lack the amino-terminal transactivation domain. The Maf family of transcription factors harbors a unique and highly conserved basic region-leucine zipper (bZIP) structure (21). The basic regions of dimeric Maf factors allows them to recognize a palindromic sequence referred to as the Maf Recognition Element (MARE). This sequence is composed of a 7-bp TPA-Responsive Element (TRE) or a 8-bp cyclic AMP-Responsive Element (CRE) core region and a TGC flanking sequence bound by the Extended Homology Region (EHR), exclusively found in Maf proteins (22) (**Figure 1**). This long recognition sequence thus distinguishes the Maf protein

family from other AP-1 family members and contributes to the important functions of the Maf proteins (23).

Thanks to their leucine zipper domain, Maf proteins can form homo- and heterodimers with other compatible bZIP proteins, such as Jun and Fos (24, 25). Maf proteins can also interact with other non-bZIP proteins including specific transcription factors, such as Sox family members (11).

Three isoforms exist for human c-Maf: a short form (373 amino acids), a medium form (383 amino acids), and a long form (30 amino acids more than the short form) of 38.5, 39.6, and 42 kDa respectively. In mice, only two isoforms have been reported: the long form (380 amino acids), called MAF-201, and the short form (370 amino acids). So far, distinct c-Maf products have not shown functional differences, but a potential functional specification cannot be excluded.

c-Maf is located on the chromosome 16q23.2 in humans and on chromosome 8 in mice (26, 27). c-Maf is translocated in 5–10% and/or overexpressed in 50% of multiple myelomas (MM) (28, 29). c-Maf overexpression in MM drives cyclin D2, integrin β7 and ARK5 expression and leads to proliferation, adhesion to bone marrow stroma cells, invasion, and migration of plasma cells (30). c-Maf is also highly expressed in over half of the angioimmunoblastic T-cell lymphomas (AITL) (30, 31). Transgenic overexpression of c-Maf in T cells regulates the same gene expression set as in plasma cells and induces T-cell lymphoma development in mice (30), therefore indicating that c-Maf is a bona fide oncogene contributing to the progression of hematological malignancies. c-Maf is also expressed by other cancers, such as renal or head and neck cancer, yet its expression is not systematically correlated with a bad prognosis (32).

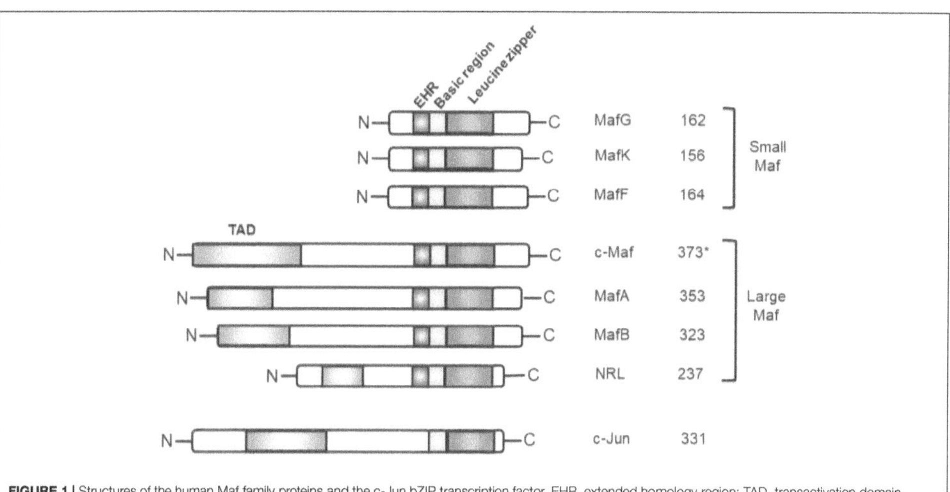

FIGURE 1 | Structures of the human Maf family proteins and the c-Jun bZIP transcription factor. EHR, extended homology region; TAD, transactivation domain. *The short isoform (isoform 1) of human c-Maf is represented.

INDUCTION OF c-MAF IN T CELLS

c-Maf expression and activity are regulated at transcriptional, post-transcriptional, as well as post-translational levels. Transcription factors and RNA-mediated silencing control the amount of Maf transcripts, while phosphorylation and SUMOylation modify the activity, sub-cellular localization, and half-life of the protein.

In T cells, antigenic stimuli that modulate the stability of the *Maf* encoding mRNAs and/or the c-Maf protein may also induce the expression of transcriptional activators, or may cooperate with independent transcriptional stimuli, such as cytokine-driven STAT factors to induce c-Maf transcription. The selective use of those pathways by different stimuli and in distinct cell populations provides the potential for tailoring c-Maf expression to different circumstances.

Transcriptional Regulation

In T cells, TCR stimulation induces the transcription of the *Maf* gene. However, distinct additional stimuli are required to sustain the expression of *Maf*, such as co-stimulatory signals (33) or the presence of cytokines, including IL-4 (34), IL-6 (35), TGF-β (36), and IL-27 (37) (**Figure 2**).

The highest levels of *Maf* transcripts can be detected in Th17 and Tfh cells. During Th17 cell polarization, both TGF-β and IL-6 are required for maximum induction of c-Maf, which in turn depends on STAT3 phosphorylation, but not STAT1 or STAT6 activation (35, 36, 38, 39). Moreover, after IL-6 stimulation,

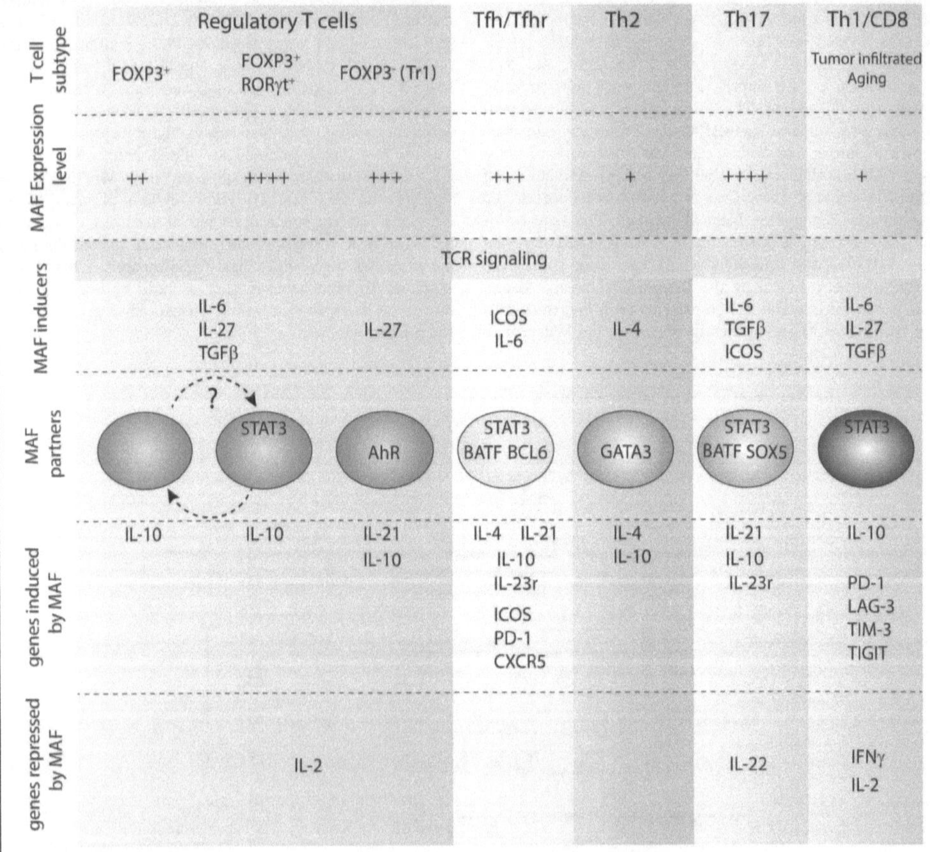

FIGURE 2 | Role of c-Maf in various T cell subtypes. Level of expression of c-Maf, signals regulating its expression, partners of c-Maf and c-Maf target genes in each indicated T cell subtype are shown.

STAT3 binds to the promoter region of *Maf* in CD4 T cells and transactivates *Maf* in a luciferase reporter gene assay (35), thus positioning STAT3 as an important STAT transcription factor for c-Maf expression in T cells.

c-Maf expression was initially thought to rely on the IL-4/STAT6 signaling pathway during Th2 cell differentiation, as ectopic expression of activated STAT6 in Th1 cells promoted c-Maf expression, along with Th2-specific cytokines and GATA3 expression (34). However, introducing GATA3 into STAT6-deficient T cells restored c-Maf expression, therefore suggesting an indirect role of STAT6 in c-Maf induction during Th2 differentiation (40). Of note, the IL-6/STAT3 signaling pathway is central to c-Maf expression during Th2 cell development (35, 41). In particular, Th2 cells express activated forms of STAT3 downstream of a STAT6-signaling pathway. Ablation of STAT3 in developing Th2 cells does not preclude GATA3 and IL-4 expression but selectively impairs c-Maf expression (41). This STAT6-to-STAT3 signaling pathway thus reconciles previous contradictory results concerning the role of STAT6 in c-Maf induction and further supports an indirect role of STAT6 in c-Maf expression.

The inducible co-stimulator (ICOS), expressed by activated T cells, promotes expression of c-Maf in murine Th2 cells and in both mouse and human Th17 cells, although the molecular mechanisms beyond this induction are still ill-defined (33, 42–44).

IL-27, a member of the IL-12/IL-23 heterodimeric family of cytokines produced by APCs, is also a potent inducer of c-Maf during Tr1 cell differentiation (45). Interestingly, IL-27 signals through STAT1/STAT3 has been shown to up-regulate ICOS expression, thus activating two independent pathways that might up-regulate c-Maf.

Prostaglandin E2 (PGE2), a pro-inflammatory lipid mediator abundant at inflammatory sites, has recently been shown to inhibit c-Maf expression in developing Tr1 cells (46). PGE2 did not affect STAT1/3 activation and its inhibitory effect was mediated through the EP4 prostaglandin receptor and cAMP signaling (46).

The expression of a c-Maf specific intergenic long non-coding RNA, called linc-MAF-4, in Th1 cells inhibits *Maf* transcription through the recruitment and activation of chromatin-modifying complexes, including the PCR2-associated histone methyltransferase, enhancer of zeste homolog 2 (EZH2), and the lysine-specific histone demethylase 1A (LSD1) (47). Up-regulation of linc-MAF-4 in human CD4 T cells is directly involved in the down-regulation of *Maf* expression and correlates with encephalitogenic Th cell differentiation and annual relapse rate in patients with multiple sclerosis (48).

Post-transcriptional Regulation

c-Maf expression is tightly regulated by small non-coding microRNAs (miRNAs). In particular, *Maf* contains phylogenetically conserved miR-155 seed matches in its 3′-UTR. Luciferase reporter experiments confirmed that the c-Maf 3′-UTR is a direct target of miR-155 (49). MiR-155 is strongly expressed in activated T cells and genetic invalidation of miR-155 led to increased levels of c-Maf in T cell lines, thus positioning

miR-155 as a major regulator of c-Maf expression *in vivo* (49). The role of miR-155 in suppressing c-Maf expression has been further extended to microglia cells during the response to CNS ischemia (50). c-Maf is also targeted by miR-143 and miR-365 in macrophages (51), and by miR-1290 in laryngeal carcinomas (52). However, expression of those miRNAs has not yet been reported in T lymphocytes.

Post-translational Control of the Biological Activity of c-Maf

Upon TCR activation, the CARMA1-dependent activation of the IKK complex results in the phosphorylation of the N-terminus part of c-Maf, which is required for nuclear translocation and binding to the promoter of target genes (53). In particular, the T cell-specific deficiency of either CARMA1 or its substrate IKKβ strongly reduced the DNA binding activity of c-Maf without affecting c-Maf abundance. This IKK-mediated activation of c-Maf is independent of NF-kB activation (53). Loss of CARMA1/IKK signaling resulted only in a partial decrease of c-Maf phosphorylation, suggesting that c-Maf might be phosphorylated by multiple kinases. c-Maf is phosphorylated by the Ser/Thr glycogen synthase kinase 3β (GSK3) in human multiple myeloma cell lines and in the lens, leading to protein stabilization (54, 55). However, whether GSK3 exerts a similar role in T cells is difficult to evaluate as GSK3 inhibition increases expression of c-Maf in this context (56).

Tyrosine phosphorylation of c-Maf is also critical for its recruitment to the IL-4 and IL-21 promoters and for optimal cytokine production. Phosphorylation of c-Maf on tyrosine residues has been shown to be positively and negatively regulated by the TEC tyrosine kinase and the PTPN22 tyrosine phosphatase, respectively (57).

SUMOylation of c-Maf at the lysine 33 residue reduces its ability to bind the *Il4* promoter and decreases the transactivating activity of c-Maf in a luciferase reporter assay (58, 59). In addition, a recent report indicated that c-Maf SUMOylation is negatively correlated with *Il21* expression in CD4 T cells from diabetogenic NOD mice (60). Furthermore, transgenic expression of a SUMO-defective c-Maf selectively inhibited recruitment of Daxx/HDAC2 to the *Il21* promoter and enhanced histone acetylation mediated by CREB-binding protein (CBP) and p300. Thus, the SUMOylation status of c-Maf has a stronger regulatory effect on IL-21 than the level of c-Maf expression, through regulation of epigenetic mechanisms (60).

ROLES OF c-MAF IN T HELPER CELLS

Regulation of IL-10 Secretion in Multiple T Cell Subsets

Multiple roles have been attributed to c-Maf in distinct T cell subsets (**Figure 2**), revealing context-specific effects of this transcription factor. However, c-Maf positively regulates *Il10* expression in virtually all immune cells, including T cells, B cells, macrophages, and dendritic cells (36, 37, 61–64), suggesting a common regulatory function beyond distinct T cell subset-specific roles.

IL-10 is an essential anti-inflammatory cytokine that plays important roles as a negative regulator of immune responses to foreign or self-antigens and prevents excessive inflammation during the course of infection [reviewed in (65–67)].

Exploring the role of c-Maf in three different disease models, each characterized by the predominant activity of a different T helper cell subset [malaria—Th1 cells; allergy to house dust mite—Th2 cells; experimental autoimmune encephalitis (EAE)—Th17 cells], Gabryšová et al. recently reported that *Il10* expression was significantly lower in the absence of c-Maf, in T helper cells across all three diseases (67). The combined evidence of open chromatin (ATAC-seq analysis) coincident with binding of c-Maf to the *Il10* locus (ChIP-seq analysis) confirmed c-Maf as a direct positive regulator of *Il10 in vivo* in distinct Th cell subsets (67).

c-Maf binds to consensus MARE motifs in the *Il10* promoter (36, 37). Although c-Maf can transactivate *Il10* by itself to some extent, c-Maf alone is not sufficient to induce optimal *Il10* expression in T cells (36, 37). Robust IL-10 expression requires interaction with additional transcriptional regulators that vary among T cell subsets. c-Maf cooperates with the aryl hydrocarbon receptor (AhR) to induce IL-10 in regulatory type 1 (Tr1) cells (37). AhR expression is mainly driven by TGF-β (68) and is not expressed in Th1 cells, in which fine-tuning IL-10 expression mostly relies on the interaction of c-Maf with Blimp-1 (64). IL-10 expression in Th2 cells relies on transcription factors STAT6, GATA3, and IRF4 (69, 70) but whether these factors interact directly with c-Maf awaits further investigation.

Thus, the activity of c-Maf on the *Il10* enhancer might not only depend on the accessibility of its motif but also on the nature of the other transcription factors that co-bind to that enhancer. In other words, *Il10* expression in Th cells relies on several transcriptional programs that, together with c-Maf, are able to integrate various signals from the environment in order to fine-tune this critical immunosuppressive cytokine.

In addition to a direct positive transcriptional regulation of *Il10* expression, c-Maf also provides a common mechanism for a negative regulation of IL-2 signaling *in vivo* in models of Th1, Th2, and Th17 responses (67).

However, although c-Maf uses common mechanisms of gene regulation in distinct cell subsets, the net outcome of c-Maf-deficiency is different in each cell type, thus indicating that c-Maf has context-specific effects on the immune response, over and above its effects on IL-10 and IL-2 signaling.

Context-Specific Effects of c-Maf on Th Cell Subset Function

Th1/Th2 Cells

The cross-regulation between Th1 and Th2 cells is mediated, in part, by the transcription factors that they express. c-Maf was first described as a Th2-specific gene that induces *Il4* gene transcription via direct binding to the *Il4* but not the *Il5* or *Il13* locus (71). Transcription factors GATA3, STAT6, and NFAT can synergize with c-Maf to regulate IL-4 expression in Th2 cells (72–74). Moreover, overexpression of c-Maf skews the immune response toward a Th2 response (75).

Although these pioneer studies concluded to a pro-Th2 role of c-Maf, normal levels of IL-13 and IgE were observed in c-Maf-deficient mice (72). Contrary to a pro-Th2 effect of c-Maf on the immune response, increased Th2 lung pathology, associated with higher numbers of eosinophils in bronchoalveolar lavage fluids, was observed in the HDM allergy model despite a decreased expression of *Il4* in T-cell specific c-Maf deficient mice (67). Of note, cells producing both IL-4 and IL-10, but not IL-4+ IL-10− Th cells, were lost in this allergy model, in keeping with increased pathology. Thus, although c-Maf can activate the *Il4* promoter, its net effect over the Th2 inflammatory response is mainly inhibitory.

In contrast with data obtained in naïve Th cells, ectopic expression of c-Maf in mature Th1 cells did not grant them the ability to produce IL-4, but did decrease their production of IFN-γ (75). In a recent study, chronically activated Th1 cells that were cultured with IL-4-producing Th2 cells up-regulate *Maf* expression (76). These cells were shown to down-regulate *Ifng* expression and express a dampened Th1 encephalitogenic cell capacity *in vivo*, despite normal expression of T-bet. Blockade of IL-4R signaling inhibited c-Maf expression in Th1 cells, suggesting that c-Maf may act downstream of IL-4R signaling to inhibit IFN-γ production in chronically activated Th1 cells (76). Thus, IL-4-driven expression of c-Maf in Th1 cells contributes to a transcriptional regulation program dampening their pathogenic immune response through altered cytokine profile.

In the malaria model, c-Maf-depletion led to greater acute-phase pathology, associated with enhanced expression of *Tbx21* and production of IFN-γ (67). This suggests a wider role of c-Maf, i.e., dampening expression of the master transcription factor T-bet, in this experimental context. However, no direct binding of c-Maf to the *Tbx21* locus was observed in the malaria model, indicating that c-Maf could regulate the expression of *Tbx21* through indirect mechanisms. Gabryšová et al. showed that the chromatin landscape of Th1 cells is remodeled by c-Maf. They identified a strong enrichment of the Runx transcription factor-binding site in the remodeled loci and further observed increased Runx expression in c-Maf-deficient Th1 cells. Given the reported effects of Runx factors on IFN-γ production (77), it is tempting to postulate that c-Maf dampens Th1 cell differentiation at least partially via repression of Runx expression. Yet functional validation of this c-Maf/Runx control over Th1 pathology still awaits experimental testing.

Those c-Maf mediated changes in the chromatin landscape were not observed in the context of Th2 or Th17 cell pathologies, again showing that the role of c-Maf varies widely depending on T helper cell type.

Follicular Helper T Cells

Follicular helper T cells (Tfh) are key regulators of T cell-dependent long-term humoral immunity (78). Tfh cells express BCL6, a transcriptional repressor considered as the critical master regulator of Tfh cell development *in vivo* (79, 80), and also constitute the major source of IL-21, a cytokine necessary for IgG class-switch recombination and antibody affinity maturation (81).

Using retroviral ectopic expression of c-Maf or BCL6 in *in vitro*-derived human Tfh cells, Kroenke et al. first reported that c-Maf and BCL6 regulate distinct features of Tfh cell functions, with BCL6 required for Tfh cell development and c-Maf for promoting IL-21 secretion (82). However, recent data have shown that c-Maf is expressed early during Tfh cell differentiation and is critical for Tfh cell development *in vivo* (83). This is in agreement with the finding that Tfh cell differentiation strongly relies on ICOSL/ICOS and IL-6/STAT3 signaling, two pathways known to induce the expression of c-Maf (35, 84–86).

The relative roles of c-Maf, BCL6, and other transcription factors in initiating and maintaining Tfh cell are still poorly defined. As c-Maf-bound genes in Tfh cells hardly correlate with genes bound by BCL6 or Ascl2 (87, 88), it is tempting to speculate that cooperation between c-Maf and BCL6/Ascl2 is required to reach complete Tfh cell fate through orchestration of distinct sets of genes. Although c-Maf does not regulate *Bcl6* transcription, a defect in BCL6 expression in CD4 T cells was observed in the absence of c-Maf, thus suggesting that c-Maf could contribute to BCL6 expression in developing Tfh cells (83).

Beside its role in Tfh cell development, c-Maf is also required for adequate IL-4 and IL-21 production through transactivation of the *Il4* and *Il21* promoters (82, 89, 90). In particular, Sahoo et al. reported that c-Maf promoted IL-4 secretion in Tfh cells through both direct binding to the CNS2 region in the *Il4* locus and via induction of IRF4, thus revealing a distinct role of c-Maf in IL-4 secretion between Th2 and Tfh cell subsets (90).

Th17 Cells

c-Maf is highly expressed in Th17 cells and impacts several important aspects of their differentiation and function (33, 36, 39, 61, 64, 67, 91) (**Figure 3**). It physically associates with the transcription factor Sox5 and, together, they bind and activate the promoter of *Rorc* in conventional CD4 T cells. The Maf-Sox5 interaction thus controls Th17 development via the induction of RORγt as downstream targets of STAT3 (92). c-Maf also positively regulates certain loci in Th17 cells, including several genes known for controlling inflammation (e.g., *Il9*, *Lif*, *Il10*) (33, 37, 82, 89). c-Maf binds the *Il21* promoter, inducing the production of IL-21 which subsequently sustains Th17 expansion (83) and stabilization through IL-23R expression (33). Some evidence also suggests that c-Maf-induced IL-21 secretion could trigger a positive feedback loop by activating STAT3, thus further promoting c-Maf and leading to the development of memory Th17 cells (93). Sato et al. suggested that c-Maf can also directly transactivate the *Il23r* gene which contains a MARE-like sequence (93). However, the relation between IL-23R and c-Maf is still controversial as it was recently reported that c-Maf downregulates *Il23r* expression in a subset of memory Th17 cells (91). c-Maf has been shown to act as a repressor of *Il2* in CD4 T cells, which indirectly amounts to increased Th17 differentiation in EAE models. Indeed, T cell specific c-Maf deficiency led to an improvement of the disease through increased IL-2 production and decreased Th17 differentiation (67).

c-Maf also acts as a global negative regulator of genes associated with Th17 function. This was demonstrated by an exhaustive study of the regulatory network for Th17 cell

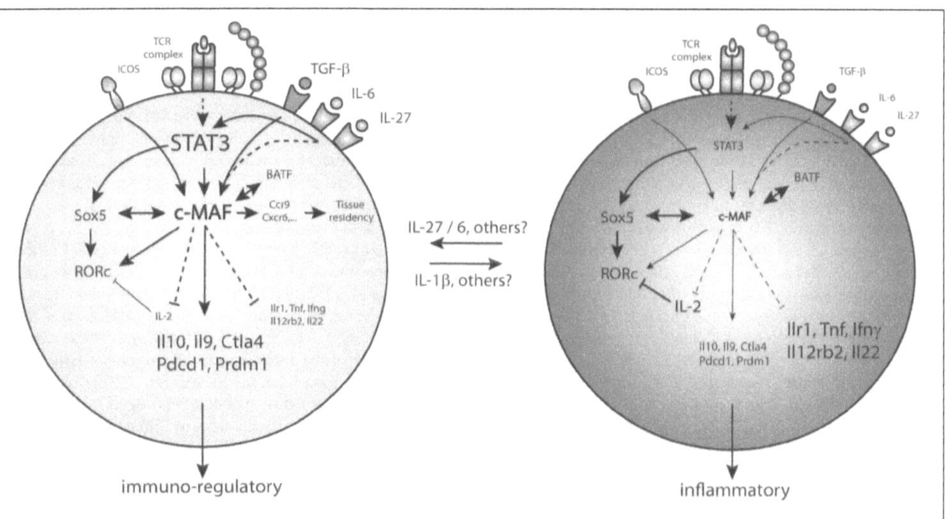

FIGURE 3 | c-Maf regulates pro/anti-inflammatory properties of Th17 cells. Under the control of STAT3, c-Maf level in Th17 cells leads to differential expression of pro-inflammatory genes (*IIr1*, *Tnfa*, *Ifng*,…) and immune-regulatory genes (*II10*, *Ctla4*, *Pdcd1*,…) in Th17 cells.

specification, where c-Maf attenuates the expression of pro-inflammatory loci (e.g., *Rora, Runx1, Il1r1, Ccr6, Tnf*) and repressing genes belonging to pathways regulated by other core transcription factors. c-Maf represses IL-22, but not IL-17A production, in a TGF-β-dependent way. It binds to the *Il22* promoter and blocks the positive transcriptional effects of other Th17 transcription factors on the *Il22* locus, including RORγt and BATF (39).

Importantly, c-Maf maintains tolerance through the regulation of IL-10, as described above. In Th17 cells, c-Maf regulates the balance between the differentiation toward inflammatory or anti-inflammatory Th17 cells (91). c-Maf controls this balance by binding to enhancers or putative enhancers already available in the general landscape of Th17 cells but not through the direct binding to promoters (91). It promotes an immuno-regulatory program (*Il10, Ctla4*) while repressing pro-inflammatory associated genes (*Ifng, Il22, Il12rb2*) (91). Interestingly, c-Maf also promotes the expression of genes associated with memory Th17 cell tissue-residency, such as *Ccr9* and *Cxcr6* (91).

The environmental cues guiding the transition between inflammatory and anti-inflammatory Th7 cells are still not clearly established. The presence of inflammatory cytokines, such as IL-1β, favors an inflammatory state, as described in Aschenbrenner et al. (91), without directly affecting the level of c-Maf in the cell. The levels of STAT3 activating cytokines, such as IL-6, IL-27, or IL-10, may have the opposite effect and push Th17 cells toward a more anti-inflammatory phenotype through c-Maf induction (36, 91). Altogether, c-Maf plays a critical role in Th17 differentiation and functions by balancing the inflammatory properties of these cells, and by doing so, it plays a critical role in the regulation of tolerance/inflammation.

Regulatory T Cells (Treg) and Gut Tolerance

Regulatory type 1 (Tr1) cells have emerged as an important subset of T cells with strong immunosuppressive properties but which does not express master transcription factor Forkhead box 3 (Foxp3), contrary to regulatory T cells (Tregs). The protective role of Tr1 cells has been shown in numerous contexts, such as autoimmunity, colitis, graft-versus-host disease, and tissue inflammation (94). The anti-inflammatory effects of Tr1 cells mainly rely on their ability to produce high amounts of IL-10. The cytokine IL-27 is known to promote the expansion and the differentiation of Tr1 cells through the induction of transcription factor c-Maf and co-stimulatory receptor ICOS. In these cells, c-Maf physically interacts with AhR to transactivate the *Il10* and *Il21* promoters (37, 45, 95). Of note, the expression of IL-21 further sustains c-Maf expression in Tr1 cells, as a feed-forward loop, thus highlighting the major role of c-Maf, both in the induction and the stabilization of the Tr1 cell fate (37, 95).

In Foxp3$^+$ Tregs, c-Maf is expressed at various levels. The fraction of c-Maf$^+$ cells found in Tregs depends on the organ studied. In the thymus, spleen, mesenteric lymph nodes or lungs, 5–40% of Foxp3$^+$ Tregs express c-Maf. Strikingly, a large proportion, ranging from 60 to 80%, of Foxp3$^+$ Tregs express c-Maf in the gut (38, 96–99). In lymphoid organs, c-Maf expression is restricted to effector Tregs (eTreg), which express high levels

of CD44 and low levels of CD62L (96, 98, 99). c-Maf expression is also induced when naïve Tregs are stimulated *in vitro* with anti-CD3 and anti-CD28 antibodies (96).

Distinct Treg populations adopt specialized phenotypes by co-expressing Foxp3 and lineage-defining transcription factors in response to tissue- or inflammatory-driven signals (100). In particular, RORγt$^+$ Tregs represent a subset of Tregs that develops in the intestinal tissue of naïve mice in response to signals arising from a complex microbiota. This subset of peripherally induced Tregs has been shown to efficiently protect from intestinal immunopathology in different colitis models (101, 102) and to mediate immunological tolerance to the gut pathobiont *Helicobacter hepaticus* (38). RORγt$^+$ Tregs express high levels of c-Maf and genetic ablation of c-Maf leads to a severe defect in the development of this subset (38, 96–99).

Numerous studies published in the last couple of years highlighted the critical role of c-Maf in the regulation of gut homeostasis. As mentioned above, c-Maf promotes IL-10 production in distinct T cell subsets. Using IL-10 reporter mice, Neumann et al. recently showed that IL-10 production by Tregs was strictly associated with c-Maf expressing Tregs (96). IL-10 is critically involved in regulating intestinal homeostasis as its total invalidation (103) or T cell specific inactivation (104–107) led to the development of a strong colitis in mice. IL-10 regulation therefore underlies a pivotal aspect of c-Maf control over gut homeostasis. Of interest, c-Maf regulates the development of RORγt$^+$ Tregs, which are the major source of IL-10 in this organ (38, 97). Thus, c-Maf might control IL-10 production in the intestine both directly, through regulation of *Il10* gene transcription, and indirectly, via the differentiation of a high IL-10-producing Treg subset.

Genetic ablation of c-Maf also impairs the differentiation of follicular regulatory T cells (Tfr) in Peyer's patches (98). Tfr cells co-express both Tfh (CXCR5, PD-1, Bcl6) and Treg (Foxp3, CTLA4) markers and represent a regulatory T cell subset that controls the activity of Tfh cells. In the intestine, Tfr cells foster microbiota diversity via the regulation of symbiotic bacteria-specific IgA affinity (108). However, the functional consequences of Tfr cell deficiency in the gut homeostasis of mice harboring c-Maf-deficient Tregs requires further investigation.

Although the control exerted by c-Maf in gut homeostasis was clearly established by different groups, the extent of the c-Maf deficiency-driven inflammation varied drastically in distinct studies. Xu et al. reported that c-Maf inactivation in Tregs (Maf$^{\Delta Treg}$) induced a colitis with the apparition of a rectal prolapse in a third of the mice (38). This phenotype was observed after infection of the animals with the pathobiont *Helicobacter hepaticus,* but was much more reduced in the absence of infection. The same mouse model did not develop spontaneous colitis, and only showed mild signs of immune cell infiltration and colon tissue destruction in other studies that were not using the pathobiont (96, 97). Surprisingly, Maf$^{\Delta Treg}$ mice were even protected from acute DSS induced colitis by an increased production of IL-17 and IL-22 (96). Lastly, c-Maf inactivation in all T cell compartments (Maf$^{\Delta T}$) resulted in a spontaneous strong onset of colitis (97).

This gradation in colitis development might be related to subtle differences in the microbiota composition of mice bred in distinct animal facilities and to a gradient of inflammation induced by the presence or not of certain pathobionts. But it might also reflect the role of c-Maf in multiple T cell subtypes in the gut (**Figure 4**).

The absence of $ROR\gamma t^+$ Tregs in the gut of $Maf^{\Delta Treg}$ mice strongly decreases the threshold for the development of colitis. However, unlike $Maf^{\Delta Treg}$ mice, $ROR\gamma t^{\Delta Treg}$ mice do not spontaneously develop colitis, suggesting that c-Maf has a more substantial role than $ROR\gamma t$ in the function of intestinal Tregs. *Il10* transcripts are still detectable in the colon of $Maf^{\Delta Treg}$ mice (97) and Imbratta et al. recently reported that Th17 cells from the colon express high levels of c-Maf and can produce IL-10. In $Maf^{\Delta T}$ mice, this additional source of IL-10 is lost and the threshold for the development of colitis decreases further. This phenotype is lost when the animals are treated with antibiotics, further confirming that the microbiota plays a critical role in the development of colitis, probably through the induction of c-Maf and the development of $ROR\gamma t^+$ Tregs in the colon (97). We can thus hypothesize that the presence of the microbiota induces a level of inflammation that triggers c-Maf expression in both intestinal Tregs and Th17 cells. c-Maf counterbalances this inflammation and promotes resistance to pathobionts through the production of IL-10 and/or the development of $ROR\gamma t^+$ Tregs (**Figure 4**).

c-Maf inactivation strongly impaired IL-10 production in both Helios positive and negative Tregs in distinct organs, showing that c-Maf also regulates IL-10 production in Tregs of thymic and peripheral origin (96). However, ablation of c-Maf in the Treg or other T cell compartments did not result in systemic autoimmune disease, nor did it disturb conventional and regulatory T cell homeostasis in lymphoid organs, and c-Maf-deficient Tregs

retained their *in vitro* suppressive capacity (38, 96, 98). Thus, although the precise role of c-Maf in the suppressive capacity of Tregs in distinct pathophysiological settings awaits further investigation, it seems that this transcription factor is a major player in colon homeostasis and that it drives immunological tolerance to gut pathobionts, thereby restraining inflammatory bowel disease.

ROLE OF c-MAF IN TUMOR INFILTRATING CD8 T CELLS

c-Maf is not expressed in CD8 T cells at steady state. The expression of c-Maf in CD8 T cells was first described in tumor infiltrating lymphocytes (TILs) obtained from a mouse melanoma model and from melanoma patient (109). Overexpression of c-Maf in CD8 T cells leads to a strong repression of IFN-γ and IL-2 production, and to an increased expression of genes associated with T cell exhaustion, i.e., a dysfunctional state of T cells observed during chronic infections or in TILs (110, 111). These genes include *Bcl6*, *Pdcd1* (PD-1), *Stat3*, and *Il10*, among other genes that are also described to be regulated by c-Maf in other T cell subtypes. When c-Maf was knocked-out in tumor specific CD8 T cells, those cells had a much higher capacity to restrain tumor growth through increased IFN-γ production and increased survival (109). Another study using B16 melanoma in mice confirmed that c-Maf was a major inducer of exhaustion associated genes, in cooperation with *Prdm1* (Blimp-1). These two transcription factors had a compensative effect for the regulation of the transcription of many inhibitory receptors, such as PD-1, TIGIT, TIM-3, or LAG-3 (112). The double knockouts for c-Maf and Blimp-1 led to a better control of B16 growth, compared to single knockouts or WT mice (112).

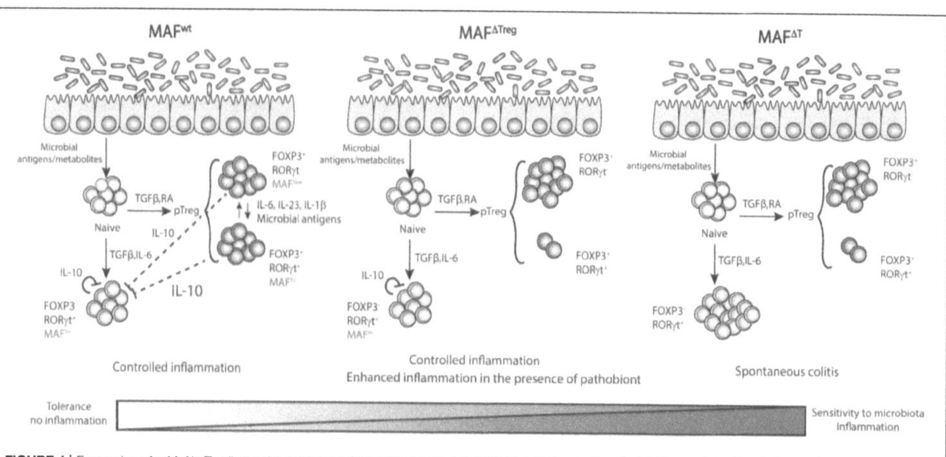

FIGURE 4 | Expression of c-Maf in T cells regulates tolerance/sensitivity to microbiota in the gut. Expression of c-Maf in regulatory T cells (Foxp3⁺ RORγt⁻ or Foxp3⁺ RORγt⁺) as well as in Th17 cells (RORγt⁺ Foxp3⁻) regulates the expression level of IL-10 in the colon and is central in the regulation of tolerance to gut microbiota.

Tc17 cells are a rather rare subtype of CD8 T cells found in gastrointestinal cancers as well as other diseases ranging from fungal infection to bacterial colonization of the skin and response to influenza (113). A study on mouse and human Tc17 cells showed that c-Maf and RORγt were both essential for the development of these cells (113). Similar to what is observed in Th cells, c-Maf expression in CD8 T cells is associated with increased tolerogenic/non-inflammatory functions, a mechanism, which is hijacked by the tumor microenvironment to favor immune escape and tumor development.

ROLE OF c-MAF IN INNATE LYMPHOID CELLS AND γδ T CELLS

Innate lymphoid cells (ILCs) are tissue-resident cells which lack antigen specificity and are preprogrammed for effector function, poising them for rapid cytokine production and for tissular immune priming. Similar to T lymphocytes, ILCs can be classified into three specialized functional subclasses, characterized by the expression of lineage-defining transcription factors and effector cytokines (114). ILC1s are defined by T-bet (*Tbx21*) and IFN-γ production. ILC2s are characterized by their high GATA3 expression and IL-5 and IL-13 secretion. ILC3s are mainly located in the intestinal lamina propria and are defined by RORγt expression. They play a crucial role in maintaining epithelial barrier integrity by expressing large amounts of IL-22 and IL-17A (115). While c-Maf has been shown to be expressed early in ILC specification (116), it is preferentially expressed in intestinal ILC3s, particularly in CCR6$^-$ NKp46$^+$ cells (17, 117).

ILCs show functional and phenotypic plasticity in response to the environment (118). This is exemplified by the ability of ILC3s to upregulate T-bet, lose RORγt expression, and convert to IFN-γ-producing ILC1s, also called "ex-ILC3s," which are involved in intestinal pathology (119–121). Recent analysis of ILC transcriptional regulatory networks showed that c-Maf regulates the ILC3/ILC1 balance in the intestine (18). c-Maf directly promotes ILC3 identity by upregulating canonical type 3 c-Maf signature genes, common to NKp46$^+$ ILC3s, CCR6$^+$ ILC3s, IL-17-producing γδ T cells, and RORγt$^+$ Tregs (17). Moreover, c-Maf indirectly sustains ILC3 identity by inducing CD127 expression, which strengthens the IL-7-dependent promotion of RORγt expression in ILCs (17). In turn, c-Maf directly represses ILC1 conversion by binding to the *Tbx21* locus and inhibiting T-bet expression, and by restraining the type 1 chromatin accessibility landscape (17). Interestingly, this direct binding of c-Maf at the *Tbx21* locus was not observed in Th1 cells and seems to be ILC-specific (67).

IL-17-producing RORγt$^+$ γδ T (Tγδ17) cells are innate-like γδ T cells, functionally preprogrammed to produce IL-17, positioning them as the main IL-17-producer in the steady state intestine and in various inflammatory diseases (122, 123). A recent study showed that the collaboration between c-Maf and RORγt is required for Tγδ17 induction and maintenance (19). In a similar dynamic to that observed in ILC3s, c-Maf sustains Tγδ17 differentiation both by supporting activating locus modifications and by antagonizing TCF1 accessibility, a negative regulator of Tγδ17 differentiation.

c-Maf therefore emerges as a global type 3 innate cell insulator, which sustains the expression type 3 genes, while constraining the acquisition of a type 1 phenotype by directly repressing type 1 genes, such as *Tbx21* and *Tcf7*, and by restraining chromatin accessibility. This feature distinguishes type 3 innate-like lymphoid cells from the adaptative Th17 cell subset, in which lineage-defining transcription factor expression and chromatin remodeling are regulated indirectly by c-Maf through negative regulation of the IL-2 signaling (67). c-Maf thus seems to occupy a higher position in the regulatory network controlling type 3 innate cells compared to that of Th17 cells.

PERSPECTIVES: MANIPULATING c-MAF EXPRESSION TO TREAT PATIENTS?

In all T cell subtypes, c-Maf plays an important role in regulating tolerance and homeostasis. Targeting c-Maf expression or function could be a good opportunity to enhance or repress a given immune response. As c-Maf is overexpressed in 50% of multiple myeloma, several laboratories developed strategies to find some potent inhibitors, which could be used to enhance the immune response, in an anti-cancer setting for instance. USP5 is important to regulate the degradation of c-Maf (124). Molecules targeting this pathway, such as Mebendazole, showed some potent anti-myeloma activity by inhibiting the USP5/c-Maf axis (125). Another possible way to inhibit c-Maf activity is by preventing its dimerization. This could be achieved by the design of specific peptide able to destabilize homodimers of c-Maf (126). An older study described that glucocorticoids are potential inhibitors of c-Maf in multiple myeloma (127). However, using these compounds to manipulate the immune system is rather inappropriate, as glucocorticoids strongly dampen it. Lowering c-Maf expression at the cellular level before T cell transfer is also one possibility, with the use of miRNA or similar technologies.

Increasing c-Maf expression in Tregs would be of crucial interest to treat autoimmune diseases, especially in the gut where c-Maf is important in several T cell subtypes. The c-Maf expressing RORγt$^+$ Foxp3$^+$ Treg population is also present in human colon but do not show any variation in proportion of total CD4 T cells when comparing colon from patients with Crohn's disease and healthy tissues (128). How to manipulate and increase c-Maf expression and RORγt$^+$ Tregs remains a challenge that requires further studies. The microbiota appears to play a central role in the regulation of this regulatory population (97, 101, 128, 129) and manipulating c-Maf expression might have dramatic impact on microbiota equilibrium. Thus, a better understanding of the c-Maf-driven Treg/Tfr/microbiota interplay would be required prior to envisage c-Maf-targeted immunotherapy.

AUTHOR CONTRIBUTIONS

CI, HH, FA, and GV wrote the text and designed the figures.

FUNDING

This work was supported by the University of Lausanne, grants from the Swiss Cancer League (GV: 3679-08-2015), the Max Cloëtta Foundation (GV), the Swiss National Foundation (GV 310030_182680), the European Regional Development Fund (ERDF), the Walloon Region (Wallonia-Biomed portfolio, 411132-957270), and the Fonds Jean Brachet, and a research credit from the National Fund for Scientific Research, FNRS, Belgium. FA is a Research Associate at the FNRS, HH was supported by a Belgian FRIA fellowship.

ACKNOWLEDGMENTS

We thank all the members of our labs for their great support.

REFERENCES

1. Enzinger FM, Shiraki, M. Musculo-aponeurotic fibromatosis of the shoulder girdle (extra-abdominal desmoid). Analysis of thirty cases followed up for ten or more years. *Cancer.* (1967) 20:1131–40.

2. Nishizawa M, Kataoka K, Goto N, Fujiwara KT, Kawai S. v-maf, a viral oncogene that encodes a "Leucine Zipper" motif. *Proc Natl Acad Sci USA.* (1989) 86:7711–5. doi: 10.1073/pnas.86.20.7711

3. Kawai S, Goto N, Kataoka K, Saegusa T, Shinno-Kohno H, Nishizawa M. Isolation of the avian transforming retrovirus, AS42, carrying the v-maf oncogene and initial characterization of its gene product. *Virology.* (1992) 188:778–84. doi: 10.1016/0042-6822(92)90532-T

4. Moussa WA, Tadros MD, Mekhael KG, Darwish AE, Shakir AH, El-Rehim EA. Some simple methods of home processing and their implication with weaning foods. *Nahrung.* (1992) 36:26–33. doi: 10.1002/food.199203 60106

5. Kawauchi S, Takahashi S, Nakajima O, Ogino H, Morita M, Nishizawa M, et al. Regulation of lens fiber cell differentiation by transcription factor c-Maf. *J Biol Chem.* (1999) 274:19254–60. doi: 10.1074/jbc.274.27. 19254

6. Ring BZ, Cordes SP, Overbeek PA, Barsh GS. Regulation of mouse lens fiber cell development and differentiation by the Maf gene. *Development.* (2000) 127:307–17.

7. Kim JI, Li T, Ho IC, Grusby MJ, Glimcher LH. Requirement for the c-Maf transcription factor in crystallin gene regulation and lens development. *Proc Natl Acad Sci USA.* (1999) 96:3781–5. doi: 10.1073/pnas.96. 7.3781

8. Wende H, Lechner SG, Cheret C, Bourane S, Kolanczyk ME, Pattyn A, et al. The transcription factor c-Maf controls touch receptor development and function. *Science.* (2012) 335:1373–6. doi: 10.1126/science. 1214314

9. Wende H, Lechner SG, Birchmeier C. The transcription factor c-Maf in sensory neuron development. *Transcription.* (2012) 3:285–9. doi: 10.4161/trns.21809

10. Imaki J, Tsuchiya K, Mishima T, Onodera H, Kim JI, Yoshida K, et al. Developmental contribution of c-maf in the kidney: distribution and developmental study of c-maf mRNA in normal mice kidney and histological study of c-maf knockout mice kidney and liver. *Biochem Biophys Res Commun.* (2004) 320:1323–7. doi: 10.1016/j.bbrc.2004. 05.222

11. Huang W, Lu N, Eberspaecher H, De Crombrugghe B. A new long form of c-Maf cooperates with Sox9 to activate the type II collagen gene. *J Biol Chem.* (2002) 277:50668–75. doi: 10.1074/jbc.M206544200

12. Hong E, di Cesare PE, Haudenschild DR. Role of c-maf in chondrocyte differentiation: a review. *Cartilage.* (2011) 2:27–35. doi: 10.1177/1947603510377464

13. MacLean HE, Kim JI, Glimcher MJ, Wang J, Kronenberg HM, Glimcher LH. Absence of transcription factor c-maf causes abnormal terminal differentiation of hypertrophic chondrocytes during endochondral bone development. *Dev Biol.* (2003) 262:51–63. doi: 10.1016/S0012-1606(03)00324-5

14. Kusakabe M, Hasegawa K, Hamada M, Nakamura M, Ohsumi T, Suzuki H, et al. c-Maf plays a crucial role for the definitive erythropoiesis that accompanies erythroblastic island formation in the fetal liver. *Blood.* (2011) 118:1374–85. doi: 10.1182/blood-2010-08-300400

15. Tsuchiya M, Taniguchi S, Yasuda K, Nitta K, Maeda A, Shigemoto M, et al. Potential roles of large mafs in cell lineages and developing pancreas. *Pancreas.* (2006) 32:408–16. doi: 10.1097/01.mpa.0000220867.64 787.99

16. Kataoka K, Shioda S, Ando K, Sakagami K, Handa H, Yasuda K. Differentially expressed Maf family transcription factors, c-Maf and MafA, activate glucagon and insulin gene expression in pancreatic islet alpha- and beta-cells. *J Mol Endocrinol.* (2004) 32:9–20. doi: 10.1677/jme.0.0320009

17. Parker ME, Barrera A, Wheaton JD, Zuberbuehler MK, Allan DSJ, Carlyle JR, et al. c-Maf regulates the plasticity of group 3 innate lymphoid cells by restraining the type 1 program. *J Exp Med.* (2020) 217:e20191030. doi: 10.1084/jem.20191030

18. Pokrovskii M, Hall JA, Ochayon DE, Yi R, Chaimowitz NS, Seelamneni H, et al. Characterization of transcriptional regulatory networks that promote and restrict identities and functions of intestinal innate lymphoid cells. *Immunity.* (2019) 51:185–97.e186. doi: 10.1016/j.immuni.2019.06.001

19. Zuberbuehler MK, Parker ME, Wheaton JD, Espinosa JR, Salzler HR, Park E, et al. The transcription factor c-Maf is essential for the commitment of IL-17-producing gammadelta T cells. *Nat Immunol.* (2019) 20:73–85. doi: 10.1038/s41590-018-0274-0

20. Liu M, Zhao X, Ma Y, Zhou Y, Deng M, Ma Y. Transcription factor c-Maf is essential for IL-10 gene expression in B cells. *Scand J Immunol.* (2018) 88:e12701. doi: 10.1111/sji.12701

21. Vinson C, Acharya A, Taparowsky EJ. Deciphering B-ZIP transcription factor interactions *in vitro* and *in vivo*. *Biochim Biophys Acta.* (2006) 1759:4–12. doi: 10.1016/j.bbaexp.2005.12.005

22. Kataoka K. Multiple mechanisms and functions of Maf transcription factors in the regulation of tissue-specific genes. *J Biochem.* (2007) 141:775–81. doi: 10.1093/jb/mvm105

23. Kerppola TK, Curran T. A conserved region adjacent to the basic domain is required for recognition of an extended DNA binding site by Maf/Nrl family proteins. *Oncogene.* (1994) 9:3149–58.

24. Kerppola TK, Curran T. Maf and Nrl can bind to AP-1 sites and form heterodimers with Fos and Jun. *Oncogene.* (1994) 9:675–84.

25. Kataoka K, Fujiwara KT, Noda M, Nishizawa M. MafB, a new Maf family transcription activator that can associate with Maf and Fos but not with Jun. *Mol Cell Biol.* (1994) 14:7581–91. doi: 10.1128/MCB.14.11.7581

26. Yoshida MC, Nishizawa M, Kataoka K, Goto N, Fujiwara KT, Kawai S. Localization of the human MAF protooncogene on chromosome 16 to bands q22-q23. (Abstract). *Cytogenet Cell Genet.* (1991) 58:2003.

27. Jamieson RV. Domain disruption and mutation of the bZIP transcription factor, MAF, associated with cataract, ocular anterior segment dysgenesis and coloboma. *Hum Mol Genet.* (2002) 11:33–42. doi: 10.1093/hmg/11.1.33

28. Chesi M, Bergsagel PL, Shonukan OO, Martelli ML, Brents LA, Chen T, et al. Frequent dysregulation of the c-maf proto-oncogene at 16q23 by translocation to an Ig locus in multiple myeloma. *Blood.* (1998) 91:4457–63. doi: 10.1182/blood.V91.12.4457.412k48_4457_4463

29. Hurt EM, Wiestner A, Rosenwald A, Shaffer AL, Campo E, Grogan T, et al. Overexpression of c-maf is a frequent oncogenic event in multiple myeloma that promotes proliferation and pathological interactions with bone marrow stroma. *Cancer Cell.* (2004) 5:191–9. doi: 10.1016/S1535-6108(04)0 0019-4

30. Morito N, Yoh K, Fujioka Y, Nakano T, Shimohata H, Hashimoto Y, et al. Overexpression of c-Maf contributes to T-cell lymphoma in both mice and human. *Cancer Res.* (2006) 66:812–9. doi: 10.1158/0008-5472.CAN-05-2154

31. Murakami YI, Yatabe Y, Sakaguchi T, Sasaki E, Yamashita Y, Morito N, et al. C-Maf expression in angioimmunoblastic T-cell lymphoma. *Am J Surg Pathol.* (2007) 31:1695–702. doi: 10.1097/PAS.0b013e318054dbcf

32. Uhlen M, Zhang C, Lee S, Sjöstedt E, Fagerberg L, Bidkhori G, et al. A pathology atlas of the human cancer transcriptome. *Science.* (2017) 357:eaan2507. doi: 10.1126/science.aan2507

33. Bauquet AT, Jin H, Paterson AM, Mitsdoerffer M, Ho I-C, Sharpe AH, et al. The costimulatory molecule ICOS regulates the expression of c-Maf and IL-21 in the development of follicular T helper cells and TH-17 cells. *Nat Immunol.* (2009) 10:167–175. doi: 10.1038/ni.1690

34. Kurata H, Lee HJ, O'Garra A, Arai N. Ectopic expression of activated Stat6 induces the expression of Th2-specific cytokines and transcription factors in developing Th1 cells. *Immunity.* (1999) 11:677–88. doi: 10.1016/S1074-7613(00)80142-9

35. Yang Y, Ochando J, Yopp A, Bromberg JS, Ding Y. IL-6 plays a unique role in initiating c-Maf expression during early stage of CD4 T cell activation. *J Immunol.* (2005) 174:2720–9. doi: 10.4049/jimmunol.174.5.2720

36. Xu J, Yang Y, Qiu G, Lal G, Wu Z, Levy DE, et al. c-Maf regulates IL-10 expression during Th17 polarization. *J Immunol.* (2009) 182:6226–36. doi: 10.4049/jimmunol.0900123

37. Apetoh L, Quintana FJ, Pot C, Joller N, Xiao S, Kumar D, et al. The aryl hydrocarbon receptor interacts with c-Maf to promote the differentiation of type 1 regulatory T cells induced by IL-27. *Nat Immunol.* (2010) 11:854–61. doi: 10.1038/ni.1912

38. Xu M, Pokrovskii M, Ding Y, Yi R, Au C, Harrison OJ, et al. c-MAF-dependent regulatory T cells mediate immunological tolerance to a gut pathobiont. *Nature.* (2018) 554:373–7. doi: 10.1038/nature25500

39. Rutz S, Noubade R, Eidenschenk C, Ota N, Zeng W, Zheng Y, et al. Transcription factor c-Maf mediates the TGF-beta-dependent suppression of IL-22 production in T(H)17 cells. *Nat Immunol.* (2011) 12:1238–45. doi: 10.1038/ni.2134

40. Ouyang W, Löhning M, Gao Z, Assenmacher M, Ranganath S, Radbruch A, et al. Stat6-independent GATA-3 autoactivation directs IL-4-independent Th2 development and commitment. *Immunity.* (2000) 12:27–37. doi: 10.1016/S1074-7613(00)80156-9

41. Mari N, Hercor M, Denanglaire S, Leo O, Andris F. The capacity of Th2 lymphocytes to deliver B-cell help requires expression of the transcription factor STAT3. *Eur J Immunol.* (2013) 43:1489–98. doi: 10.1002/eji.201242938

42. Coyle AJ, Lehar S, Lloyd C, Tian J, Delaney T, Manning S, et al. The CD28-related molecule ICOS is required for effective T cell-dependent immune responses. *Immunity.* (2000) 13:95–105. doi: 10.1016/S1074-7613(00)00011-X

43. Nurieva RI, Duong J, Kishikawa H, Dianzani U, Rojo JM, Ho IC, et al. Transcriptional regulation of Th2 differentiation by inducible costimulator. *Immunity.* (2003) 18:801–11. doi: 10.1016/S1074-7613(03)00144-4

44. Paulos CM, Paulos CM, Carpenito C, Plesa G, Suhoski MM, Varela-rohena A, et al. The inducible costimulator (ICOS) is critical for the development of human T H 17 cells. (2010) 2:55ra78. doi: 10.1126/scitranslmed.3000448

45. Pot C, Jin H, Awasthi A, Liu SM, Lai C-Y, Madan R, et al. Cutting edge: IL-27 induces the transcription factor c-Maf, cytokine IL-21, and the costimulatory receptor ICOS that coordinately act together to promote differentiation of IL-10-producing Tr1 cells. *J Immunol.* (2009) 183:797–801. doi: 10.4049/jimmunol.0901233

46. Hooper KM, Kong W, Ganea D. Prostaglandin E2 inhibits Tr1 cell differentiation through suppression of c-Maf. *PLoS ONE.* (2017) 12:e0179184. doi: 10.1371/journal.pone.0179184

47. Ranzani V, Rossetti G, Panzeri I, Arrigoni A, Bonnal RJ, Curti S, et al. The long intergenic noncoding RNA landscape of human lymphocytes highlights the regulation of T cell differentiation by linc-MAF-4. *Nat Immunol.* (2015) 16:318–25. doi: 10.1038/ni.3093

48. Zhang F, Liu G, Wei C, Gao C, Hao J. Linc-MAF-4 regulates Th1/Th2 differentiation and is associated with the pathogenesis of multiple sclerosis by targeting MAF. *FASEB J.* (2017) 31:519–25. doi: 10.1096/fj.201600838R

49. Rodriguez A, Vigorito E, Clare S, Warren MV, Couttet P, Soond DR, et al. Requirement of bic/microRNA-155 for normal immune function. *Science.* (2007) 316:608–11. doi: 10.1126/science.1139253

50. Su W, Hopkins S, Nesser NK, Sopher B, Silvestroni A, Ammanuel S, et al. The p53 transcription factor modulates microglia behavior through microRNA-dependent regulation of c-Maf. *J Immunol.* (2014) 192:358–66. doi: 10.4049/jimmunol.1301397

51. Tamgue O, Gcanga L, Ozturk M, Whitehead L, Pillay S, Jacobs R, et al. Differential targeting of c-Maf, Bach-1, and Elmo-1 by microRNA-143 and microRNA-365 promotes the intracellular growth of *Mycobacterium tuberculosis* in alternatively IL-4/IL-13 activated macrophages. *Front Immunol.* (2019) 10:421. doi: 10.3389/fimmu.2019.00421

52. Janiszewska J, Szaumkessel M, Kostrzewska-Poczekaj M, Bednarek K, Paczkowska J, Jackowska J, et al. Global miRNA expression profiling identifies miR-1290 as novel potential oncomiR in laryngeal carcinoma. *PLoS ONE.* (2015) 10:e0144924. doi: 10.1371/journal.pone.0144924

53. Blonska M, Joo D, Nurieva RI, Zhao X, Chiao P, Sun SC, et al. Activation of the transcription factor c-Maf in T cells is dependent on the CARMA1-IKKbeta signaling cascade. *Sci Signal.* (2013) 6:ra110. doi: 10.1126/scisignal.2004273

54. Qiang YW, Ye S, Chen Y, Buros AF, Edmonson R, van Rhee F, et al. MAF protein mediates innate resistance to proteasome inhibition therapy in multiple myeloma. *Blood.* (2016) 128:2919–30. doi: 10.1182/blood-2016-03-706077

55. Niceta M, Stellacci E, Gripp KW, Zampino G, Kousi M, Anselmi M, et al. Mutations impairing GSK3-mediated MAF phosphorylation cause cataract, deafness, intellectual disability, seizures, and a down syndrome-like facies. *Am J Hum Genet.* (2015) 96:816–25. doi: 10.1016/j.ajhg.2015.03.001

56. Hill EV, Ng TH, Burton BR, Oakley CM, Malik K, Wraith DC. Glycogen synthase kinase-3 controls IL-10 expression in CD4(+) effector T-cell subsets through epigenetic modification of the IL-10 promoter. *Eur J Immunol.* (2015) 45:1103–15. doi: 10.1002/eji.201444661

57. Liu CC, Lai CY, Yen WF, Lin YH, Chang HH, Tai TS, et al. Reciprocal regulation of C-Maf tyrosine phosphorylation by Tec and Ptpn22. *PLoS ONE.* (2015) 10:e0127617. doi: 10.1371/journal.pone.0127617

58. Leavenworth JW, Ma X, Mo YY, Pauza ME. SUMO conjugation contributes to immune deviation in nonobese diabetic mice by suppressing c-Maf transactivation of IL-4. *J Immunol.* (2009) 183:1110–9. doi: 10.4049/jimmunol.0803671

59. Lin BS, Tsai PY, Hsieh WY, Tsao HW, Liu MW, Grenningloh R, et al. SUMOylation attenuates c-Maf-dependent IL-4 expression. *Eur J Immunol.* (2010) 40:1174–84. doi: 10.1002/eji.200939788

60. Hsu CY, Yeh LT, Fu SH, Chien MW, Liu YW, Miaw SC, et al. SUMO-defective c-Maf preferentially transactivates Il21 to exacerbate autoimmune diabetes. *J Clin Invest.* (2018) 128:3779–93. doi: 10.1172/JCI98786

61. Ciofani M, Madar A, Galan C, Sellars M, Mace K, Pauli F, et al. A validated regulatory network for Th17 cell specification. *Cell.* (2012) 151:289–303. doi: 10.1016/j.cell.2012.09.016

62. Cao S, Liu J, Song L, Ma X. The Protooncogene c-Maf is an essential transcription factor for IL-10 gene expression in macrophages. *J Immunol.* (2005) 174:3484–92. doi: 10.4049/jimmunol.174.6.3484

63. Saraiva M, Christensen JR, Veldhoen M, Murphy TL, Murphy KM, O'Garra A. Interleukin-10 production by Th1 cells requires interleukin-12-induced STAT4 transcription factor and ERK MAP kinase activation by high antigen dose. *Immunity.* (2009) 31:209–19. doi: 10.1016/j.immuni.2009.05.012

64. Neumann C, Heinrich F, Neumann K, Junghans V, Mashreghi M-F, Ahlers J, et al. Role of Blimp-1 in programing Th effector cells into IL-10 producers. *J Exp Med.* (2014) 211:1807–19. doi: 10.1084/jem.20131548

65. Ouyang W, Rutz S, Crellin NK, Valdez PA, Hymowitz SG. Regulation and functions of the IL-10 family of cytokines in inflammation and disease. *Annu Rev Immunol.* (2011) 29:71–109. doi: 10.1146/annurev-immunol-031210-101312

66. Rojas JM, Avia M, Martin V, Sevilla N. IL-10: a multifunctional cytokine in viral infections. *J Immunol Res.* (2017) 2017:6104054. doi: 10.1155/2017/6104054

67. Gabryšová L, Alvarez-Martinez M, Luisier R, Cox LS, Sodenkamp J, Hosking C, et al. C-Maf controls immune responses by regulating disease-specific gene networks and repressing IL-2 in CD4+T cells article. *Nat Immunol.* (2018) 19:497–507. doi: 10.1038/s41590-018-0083-5

68. Veldhoen M, Hirota K, Westendorf AM, Buer J, Dumoutier L, Renauld JC, et al. The aryl hydrocarbon receptor links TH17-cell-mediated autoimmunity to environmental toxins. *Nature.* (2008) 453:106–9. doi: 10.1038/nature06881

69. Chang HD, Helbig C, Tykocinski L, Kreher S, Koeck J, Niesner U, et al. Expression of IL-10 in Th memory lymphocytes is conditional on IL-12 or IL-4, unless the IL-10 gene is imprinted by GATA-3. *Eur J Immunol.* (2007) 37:807–17. doi: 10.1002/eji.200636385

70. Ahyi AN, Chang HC, Dent AL, Nutt SL, Kaplan MH. IFN regulatory factor 4 regulates the expression of a subset of Th2 cytokines. *J Immunol.* (2009) 183:1598–606. doi: 10.4049/jimmunol.0803302

71. Ho IC, Hodge MR, Rooney JW, Glimcher LH. The proto-oncogene c-maf is responsible for tissue-specific expression of interleukin-4. *Cell.* (1996) 85:973–83. doi: 10.1016/S0092-8674(00)81299-4

72. Kim JI, Ho IC, Grusby MJ, Glimcher LH. The transcription factor c-Maf controls the production of interleukin-4 but not other Th2 cytokines. *Immunity.* (1999) 10:745–51. doi: 10.1016/S1074-7613(00)80073-4

73. Hodge MR, Chun HJ, Rengarajan J, Alt A, Lieberson R, Glimcher LH. NF-AT-driven interleukin-4 transcription potentiated by NIP45. *Science.* (1996) 274:1903–5. doi: 10.1126/science.274.5294.1903

74. Li B, Tournier C, Davis RJ, Flavell RA. Regulation of IL-4 expression by the transcription factor JunB during T helper cell differentiation. *Eur Mol Biol Org J.* (1999) 18:420–32. doi: 10.1093/emboj/18.2.420

75. Ho IC, Lo D, Glimcher LH. c-maf promotes T helper cell type 2 (Th2) and attenuates Th1 differentiation by both interleukin 4-dependent and -independent mechanisms. *J Exp Med.* (1998) 188:1859–66. doi: 10.1084/jem.188.10.1859

76. Mitchell RE, Hassan M, Burton BR, Britton G, Hill EV, Verhagen J, et al. IL-4 enhances IL-10 production in Th1 cells: implications for Th1 and Th2 regulation. *Sci Rep.* (2017) 7:11315. doi: 10.1038/s41598-017-11803-y

77. Djuretic IM, Levanon D, Negreanu V, Groner Y, Rao A, Ansel KM. Transcription factors T-bet and Runx3 cooperate to activate Ifng and silence Il4 in T helper type 1 cells. *Nat Immunol.* (2007) 8:145–53. doi: 10.1038/ni1424

78. Vinuesa CG, Linterman MA, Yu D, MacLennan IC. Follicular helper T cells. *Annu Rev Immunol.* (2016) 34:335–68. doi: 10.1146/annurev-immunol-041015-055605

79. Johnston RJ, Poholek AC, DiToro D, Yusuf I, Eto D, Barnett B, et al. Bcl6 and Blimp-1 are reciprocal and antagonistic regulators of T follicular helper cell differentiation. *Science.* (2009) 325:1006–10. doi: 10.1126/science.1175870

80. Nurieva RI, Chung Y, Martinez GJ, Yang XO, Tanaka S, Matskevitch TD, et al. Bcl6 mediates the development of T follicular helper cells. *Science.* (2009) 325:1001–5. doi: 10.1126/science.1176676

81. Leopold MF, Begeman L, van Bleijswijk JD, LL IJ, Witte HJ, Grone A. Exposing the grey seal as a major predator of harbour porpoises. *Proc Biol Sci.* (2015) 282:20142429. doi: 10.1098/rspb.2014.2429

82. Kroenke MA, Eto D, Locci M, Cho M, Davidson T, Haddad EK, et al. Bcl6 and Maf cooperate to instruct human follicular helper CD4 T cell differentiation. *J Immunol.* (2012) 188:3734–44. doi: 10.4049/jimmunol.1103246

83. Andris F, Denanglaire S, Anciaux M, Hercor M, Hussein H, Leo O. The transcription factor c-Maf promotes the differentiation of follicular helper T cells. *Front Immunol.* (2017) 8:480. doi: 10.3389/fimmu.2017.00480

84. Eddahri F, Denanglaire S, Bureau F, Spolski R, Leonard WJ, Leo O, et al. Interleukin-6/STAT3 signaling regulates the ability of naive T cells to acquire B-cell help capacities. *Blood.* (2009) 113:2426–33. doi: 10.1182/blood-2008-04-154682

85. Nurieva RI, Mai XM, Forbush K, Bevan MJ, Dong C. B7h is required for T cell activation, differentiation, and effector function. *Proc Natl Acad Sci USA.* (2003) 100:14163–8. doi: 10.1073/pnas.2335041100

86. Mak TW, Shahinian A, Yoshinaga SK, Wakeham A, Boucher LM, Pintilie M, et al. Costimulation through the inducible costimulator ligand is essential for both T helper and B cell functions in T cell-dependent B cell responses. *Nat Immunol.* (2003) 4:765–72. doi: 10.1038/ni947

87. Liu X, Chen X, Zhong B, Wang A, Wang X, Chu F, et al. Transcription factor achaete-scute homologue 2 initiates follicular T-helper-cell development. *Nature.* (2014) 507:513–8. doi: 10.1038/nature12910

88. Liu X, Lu H, Chen T, Nallaparaju KC, Yan X, Tanaka S, et al. Genome-wide analysis identifies Bcl6-controlled regulatory networks during T follicular helper cell differentiation. *Cell Rep.* (2016) 14:1735–47. doi: 10.1016/j.celrep.2016.01.038

89. Hiramatsu Y, Suto A, Kashiwakuma D, Kanari H, Kagami S-i, Ikeda K, et al. c-Maf activates the promoter and enhancer of the IL-21 gene, and TGF-β inhibits c-Maf-induced IL-21 production in CD4+ T cells. *J Leukoc Biol.* (2010) 87:703–12. doi: 10.1189/jlb.0909639

90. Sahoo A, Alekseev A, Tanaka K, Obertas L, Lerman B, Haymaker C, et al. Batf is important for IL-4 expression in T follicular helper cells. *Nat Commun.* (2015) 6:7997. doi: 10.1038/ncomms8997

91. Aschenbrenner D, Foglierini M, Jarrossay D, Hu D, Weiner HL, Kuchroo VK, et al. An immunoregulatory and tissue-residency program modulated by c-MAF in human TH17 cells. *Nat Immunol.* (2018) 19:1126–36. doi: 10.1038/s41590-018-0200-5

92. Tanaka S, Suto A, Iwamoto T, Kashiwakuma D, Kagami S-i, Suzuki K, et al. Sox5 and c-Maf cooperatively induce Th17 cell differentiation via RORγt induction as downstream targets of Stat3. *J Exp Med.* (2014) 211:1857–74. doi: 10.1084/jem.20130791

93. Sato K, Miyoshi F, Yokota K, Araki Y, Asanuma Y, Akiyama Y, et al. Marked induction of c-Maf protein during Th17 cell differentiation and its implication in memory Th cell development. *J Biol Chem.* (2011) 286:14963–71. doi: 10.1074/jbc.M111.218867

94. Roncarolo MG, Gregori S, Battaglia M, Bacchetta R, Fleischhauer K, Levings MK. Interleukin-10-secreting type 1 regulatory T cells in rodents and humans. *Immunol Rev.* (2006) 212:28–50. doi: 10.1111/j.0105-2896.2006.00420.x

95. Pot C, Apetoh L, Kuchroo VK. Type 1 regulatory T cells (Tr1) in autoimmunity. *Semin Immunol.* (2011) 23:202–8. doi: 10.1016/j.smim.2011.07.005

96. Neumann C, Blume J, Roy U, Teh PP, Vasanthakumar A, Beller A, et al. c-Maf-dependent Treg cell control of intestinal TH17 cells and IgA establishes host-microbiota homeostasis. *Nat Immunol.* (2019) 20:471–81. doi: 10.1038/s41590-019-0316-2

97. Imbratta C, Leblond MM, Bouzourene H, Speiser DE, Velin D, Verdeil G. Maf deficiency in T cells dysregulates Treg - TH17 balance leading to spontaneous colitis. *Sci Rep.* (2019) 9:6135. doi: 10.1038/s41598-019-42486-2

98. Wheaton JD, Yeh CH, Ciofani M. Cutting edge: c-Maf is required for regulatory T cells to adopt RORgammat(+) and follicular phenotypes. *J Immunol.* (2017) 199:3931–6. doi: 10.4049/jimmunol.1701134

99. Hussein HDS, Van Gool F, Azouz A, Ajouaou Y, El-Khatib H, Oldenhove G, et al. Multiple environmental signaling pathways control the differentiation of RORγt-expressing regulatory T cells. *Front Immunol.* (2019) 10:3007. doi: 10.3389/fimmu.2019.03007

100. Panduro M, Benoist C, Mathis D. Tissue tregs. *Annu Rev Immunol.* (2016) 34:609–33. doi: 10.1146/annurev-immunol-032712-095948

101. Ohnmacht C, Park JH, Cording S, Wing JB, Atarashi K, Obata Y, et al. Mucosal immunology. The microbiota regulates type 2 immunity through RORgammat(+) T cells. *Science.* (2015) 349:989–93. doi: 10.1126/science.aac4263

102. Lochner M, Ohnmacht C, Presley L, Bruhns P, Si-Tahar M, Sawa S, et al. Microbiota-induced tertiary lymphoid tissues aggravate inflammatory disease in the absence of RORgamma t and LTi cells. *J Exp Med.* (2011) 208:125–34. doi: 10.1084/jem.20100052

103. Kühn R, Löhler J, Rennick D, Rajewsky K, Müller W. Interleukin-10-deficient mice develop chronic enterocolitis. *Cell.* (1993) 75:263–74. doi: 10.1016/0092-8674(93)80068-P

104. Roers A, Siewe L, Strittmatter E, Deckert M, Schlüter D, Stenzel W, et al. T cell-specific inactivation of the interleukin 10 gene in mice results in enhanced T cell responses but normal innate responses to lipopolysaccharide or skin irritation. *J Exp Med.* (2004) 200:1289–97. doi: 10.1084/jem.20041789

105. Huber S, Gagliani N, Esplugues E, O'Connor W Jr, Huber FJ, Chaudhry A, et al. Th17 cells express interleukin-10 receptor and are controlled by Foxp3(-) and Foxp3+ regulatory CD4+ T cells in an interleukin-10-dependent manner. *Immunity.* (2011) 34:554–65. doi: 10.1016/j.immuni.2011.01.020

106. Rubtsov YP, Rasmussen JP, Chi EY, Fontenot J, Castelli L, Ye X, et al. Regulatory T cell-derived interleukin-10 limits inflammation at environmental interfaces. *Immunity.* (2008) 28:546–58. doi: 10.1016/j.immuni.2008.02.017

107. Chaudhry A, Samstein RM, Treuting P, Liang Y, Pils MC, Heinrich J-m, et al. Interleukin-10 signaling in regulatory T cells is required for suppression of Th17 cell-mediated inflammation. *Immunity.* (2011) 34:566–78. doi: 10.1016/j.immuni.2011.03.018

108. Rescigno M. Tfr cells and IgA join forces to diversify the microbiota. *Immunity.* (2014) 41:9–11. doi: 10.1016/j.immuni.2014.06.012

109. Giordano M, Henin C, Maurizio J, Imbratta C, Bourdely P, Buferne M, et al. Molecular profiling of CD8 T cells in autochthonous melanoma identifies Maf as driver of exhaustion. *EMBO J.* (2015) 34:2042–58. doi: 10.15252/embj.201490786

110. Speiser DE, Ho PC, Verdeil G. Regulatory circuits of T cell function in cancer. *Nat Rev Immunol.* (2016) 16:599–611. doi: 10.1038/nri.2016.80

111. Wherry EJ. T cell exhaustion. *Nat Immunol.* (2011) 12:492–9. doi: 10.1038/ni.2035

112. Chihara N, Madi A, Kondo T, Zhang H, Acharya N, Singer M, et al. Induction and transcriptional regulation of the co-inhibitory gene module in T cells. *Nature.* (2018) 558:454–9. doi: 10.1038/s41586-018-0206-z

113. Mielke LA, Liao Y, Clemens EB, Firth MA, Duckworth B, Huang Q, et al. TCF-1 limits the formation of Tc17 cells via repression of the MAF-RORgammat axis. *J Exp Med.* (2019) 216:1682–99. doi: 10.1084/jem.20181778

114. Vivier E, Artis D, Colonna M, Diefenbach A, Di Santo JP, Eberl G, et al. Innate lymphoid cells: 10 years on. *Cell.* (2018) 174:1054–66. doi: 10.1016/j.cell.2018.07.017

115. Sawa S, Lochner M, Satoh-Takayama N, Dulauroy S, Berard M, Kleinschek M, et al. RORgammat+ innate lymphoid cells regulate intestinal homeostasis by integrating negative signals from the symbiotic microbiota. *Nat Immunol.* (2011) 12:320–6. doi: 10.1038/ni.2002

116. Harly C, Cam M, Kaye J, Bhandoola A. Development and differentiation of early innate lymphoid progenitors. *J Exp Med.* (2018) 215:249–62. doi: 10.1084/jem.20170832

117. Gury-BenAri M, Thaiss CA, Serafini N, Winter DR, Giladi A, Lara-Astiaso D, et al. The spectrum and regulatory landscape of intestinal innate lymphoid cells are shaped by the microbiome. *Cell.* (2016) 166:1231–46.e1213. doi: 10.1016/j.cell.2016.07.043

118. Colonna M. Innate lymphoid cells: diversity, plasticity, and unique functions in immunity. *Immunity.* (2018) 48:1104–17. doi: 10.1016/j.immuni.2018.05.013

119. Song C, Lee JS, Gilfillan S, Robinette ML, Newberry RD, Stappenbeck TS, et al. Unique and redundant functions of NKp46+ ILC3s in models of intestinal inflammation. *J Exp Med.* (2015) 212:1869–82. doi: 10.1084/jem.20151403

120. Buonocore S, Ahern PP, Uhlig HH, Ivanov, II, Littman DR, Maloy KJ, et al. Innate lymphoid cells drive interleukin-23-dependent innate intestinal pathology. *Nature.* (2010) 464:1371–5. doi: 10.1038/nature08949

121. Klose CS, Kiss EA, Schwierzeck V, Ebert K, Hoyler T, d'Hargues Y, et al. A T-bet gradient controls the fate and function of CCR6-RORgammat+ innate lymphoid cells. *Nature.* (2013) 494:261–5. doi: 10.1038/nature11813

122. Papotto PH, Ribot JC, Silva-Santos B. IL-17(+) gammadelta T cells as kick-starters of inflammation. *Nat Immunol.* (2017) 18:604–11. doi: 10.1038/ni.3726

123. Cua DJ, Tato CM. Innate IL-17-producing cells: the sentinels of the immune system. *Nat Rev Immunol.* (2010) 10:479–89. doi: 10.1038/nri2800

124. Wang S, Juan J, Zhang Z, Du Y, Xu Y, Tong J, et al. Inhibition of the deubiquitinase USP5 leads to c-Maf protein degradation and myeloma cell apoptosis. *Cell Death Dis.* (2017) 8:e3058. doi: 10.1038/cddis.2017.450

125. Chen XH, Xu YJ, Wang XG, Lin P, Cao BY, Zeng YY, et al. Mebendazole elicits potent antimyeloma activity by inhibiting the USP5/c-Maf axis. *Acta Pharmacol Sin.* (2019) 40:1568–77. doi: 10.1038/s41401-019-0249-1

126. Pellegrino S, Ronda L, Annoni C, Contini A, Erba E, Gelmi ML, et al. Molecular insights into dimerization inhibition of c-Maf transcription factor. *Biochim Biophys Acta.* (2014) 1844:2108–15. doi: 10.1016/j.bbapap.2014.09.003

127. Mao X, Stewart AK, Hurren R, Datti A, Zhu X, Zhu Y, et al. A chemical biology screen identifies glucocorticoids that regulate c-maf expression by increasing its proteasomal degradation through up-regulation of ubiquitin. *Blood.* (2007) 110:4047–54. doi: 10.1182/blood-2007-05-088666

128. Sefik E, Geva-Zatorsky N, Oh S, Konnikova L, Zemmour D, McGuire AM, et al. Mucosal immunology. Individual intestinal symbionts induce a distinct population of RORgamma(+) regulatory T cells. *Science.* (2015) 349:993–7. doi: 10.1126/science.aaa9420

129. Kim KS, Hong SW, Han D, Yi J, Jung J, Yang BG, et al. Dietary antigens limit mucosal immunity by inducing regulatory T cells in the small intestine. *Science.* (2016) 351:858–63. doi: 10.1126/science.aac5560

Conflict of Interest: The authors declare that the research was conducted in the absence of any commercial or financial relationships that could be construed as a potential conflict of interest.

Copyright © 2020 Imbratta, Hussein, Andris and Verdeil. This is an open-access article distributed under the terms of the Creative Commons Attribution License (CC BY). The use, distribution or reproduction in other forums is permitted, provided the original author(s) and the copyright owner(s) are credited and that the original publication in this journal is cited, in accordance with accepted academic practice. No use, distribution or reproduction is permitted which does not comply with these terms.

8.2. Annex 2 – Research article: "The transcription factor c-Maf promotes the differentiation of follicular helper T cells"

Follicular helper T (Tfh) cells promote B cell responses in germinal centers, and therefore support high-affinity antibody production. We identified a role for transcription factor c-Maf in the differentiation of Tfh cells in response to both antigen/adjuvant vaccinations and commensal intestinal bacteria. This work has been published in Frontiers in Immunology on April 27th 2017 and was titled "The transcription factor c-Maf promotes the differentiation of follicular helper T cells"[406]. I perfomed the experiments assessing Tfh differentiation in Peyer's patches (Figure 3 I-K) and analyzed these data, in the context of my master book project.

ORIGINAL RESEARCH
published: 27 April 2017
doi: 10.3389/fimmu.2017.00480

The Transcription Factor c-Maf Promotes the Differentiation of Follicular Helper T Cells

Fabienne Andris*, Sébastien Denanglaire, Maelle Anciaux, Mélanie Hercor, Hind Hussein and Oberdan Leo

Laboratoire d'Immunobiologie, Université Libre de Bruxelles, Brussels, Belgium

Follicular helper T cells (Tfh) have been identified as the primary cell subpopulation regulating B cell responses in germinal centers, thus supporting high-affinity antibody production. Among the transcription factors orchestrating Tfh cell differentiation and function, the role played by the proto-oncogene c-Maf remains poorly characterized. We report herein that selective loss of c-Maf expression in the T cell compartment results in defective development of Tfh cells in response to both antigen/adjuvant vaccinations and commensal intestinal bacteria. Accordingly, c-Maf expression in T cells was essential for the development and high-affinity antibody secretion in vaccinated animals. c-Maf was expressed early, concomitantly to BCL6, in Tfh cell precursors and found to regulate Tfh fate in a cell-autonomous fashion. Altogether, our findings reveal a novel, non-redundant, function for c-Maf in the differentiation of Tfh cells and the regulation of humoral immune responses to T-cell-dependent antigens.

Keywords: CD4+ T lymphocytes, follicular helper T cells, humoral response, c-Maf, T cell differentiation

OPEN ACCESS

Edited by:
*Barbara Fazekas De St Groth,
University of Sydney, Australia*

Reviewed by:
*Kai-Michael Toellner,
University of Birmingham, UK
F. Ronchese,
Malaghan Institute of Medical
Research, New Zealand*

***Correspondence:**
*Fabienne Andris
fandris@ulb.ac.be*

Specialty section:
*This article was submitted
to T Cell Biology,
a section of the journal
Frontiers in Immunology*

Received: 14 February 2017
Accepted: 06 April 2017
Published: 27 April 2017

Citation:
*Andris F, Denanglaire S, Anciaux M,
Hercor M, Hussein H and Leo O
(2017) The Transcription Factor c-Maf
Promotes the Differentiation of
Follicular Helper T Cells.
Front. Immunol. 8:480.
doi: 10.3389/fimmu.2017.00480*

INTRODUCTION

Follicular helper T cells (Tfh) are key regulators of germinal center (GC) formation and T cell-dependent long-term humoral immunity. They provide crucial signals to B lymphocytes in GCs and guide high-affinity, isotype-switched, antibody responses, and memory B cell development (1).

Follicular helper T cells express the transcriptional repressor BCL6, considered as the critical master regulator of their development *in vivo* (2, 3) and are the major source of IL-21, which is necessary for IgG class-switch recombination and antibody-affinity maturation (4).

The differentiation of Tfh cells is considered as a multistage process starting in the T cell zone of secondary lymphoid organs. Here, T lymphocytes engage in cognate interaction and ICOS-ICOSL signaling with dendritic cells (DCs). These signals promote expression of CXCR5, allowing Th cells to relocalize at the T–B border zone where they receive additional signals from B cells (5, 6). This second wave of interactions further stabilizes Tfh cell fate—characterized by a high expression of BCL6 and surface markers such as CXCR5, PD1, ICOS—and results in the migration toward GCs and the delivery of optimal helper signals to B cells (5–7).

This stepwise differentiation pathway results from the sequential activation of a series of transcription factors regulating distinct phases of the Tfh developmental program. Within this intricate Tfh-associated transcriptional network, Ascl2 and BCL6 represent master regulators initiating Tfh cell development by inducing the expression of key Tfh-associated genes while inhibiting the expression of other, non-Tfh, helper cell subset signature genes (2, 3, 8, 9).

The transcription factor c-Maf, belonging to the AP-1 family of basic region/leucine zipper factor, is highly expressed by mature Tfh cells, and is thought to mainly regulate the expression of cytokines able to promote B cell proliferation and differentiation. Indeed, c-Maf is expressed downstream of Batf and ICOS signaling and has been shown to transactivate IL-4 and IL-21 promoters (10–12). In particular, Sahoo et al. recently reported that c-Maf promotes IL-4 secretion in Tfh cells through both direct binding to the CNS2 region in the *il4* locus and via induction of IRF4, thus revealing a distinct role of c-Maf in IL-4 secretion between Th2 and Tfh cell subsets (12).

Collectively, the available literature posits c-Maf as an important regulator of cytokine production in Tfh cells, thus acting at a later stage of the Tfh developmental program (1, 10, 12). To directly evaluate the putative role of c-Maf in the generation and regulation of Tfh activity, we have characterized the immune response of mice selectively lacking c-Maf expression in the T cell compartment. In contrast to our expectations, T cells lacking c-Maf expression failed to acquire expression of key Tfh markers (such as BCL6, CXCR5, and PD1), indicating an important, and non-redundant role for c-Maf in the initiation of Tfh cell development. Accordingly, mice lacking c-Maf in the T cell compartment displayed reduced secretion of high-affinity antibodies. Our data thus uncover a major and unsuspected role for c-Maf in regulating Tfh cell development and T-cell-dependent humoral responses.

MATERIALS AND METHODS

Mice and Immunization

C57BL/6 mice were purchased from Envigo (Horst, The Netherlands). c-Maf-flox mice (13) were kindly provided by Dr. Carmen Birchmeier (Max Delbrück Center for Molecular Medicine, Berlin, Germany) and were back-crossed for nine generations to C57BL/6 in our animal facility before breeding with CD4-CRE mice (14), provided by Dr. Geert Van Loo (University of Gent, Gent, Belgium) to generate T-cell compartment-specific c-Maf-deficient mice (c-Maf^{KO-T} mice). CD3ε-KO mice were from EMMA (CDTA, Orleans, France).

All mice were used at 6–12 weeks of age.

Mice were immunized by injecting 10 µg keyhole limpet hemocyanin (KLH, Calbiochem) in foot pads (f.p.) along with Alum (1 mg/f.p., Thermo Fisher Scientific, Rockford, IL, USA) or IFA (sigma; 25 µL/f.p.) supplemented with LPS (*Escherichia coli* serotype 0111:B5, Calbiochem; 5 µg/f.p.). In some experiments, mice were immunized intra-peritoneally (i.p.) with 75 µg nitrophenyl-KLH (NP$_{25}$-KLH, Biosearch Technologies, Novato, CA, USA) and 1 mg of Imject Alum. When indicated, mice were further boosted on day 14 by a second immunization with NP-KLH in saline.

Differentiation of BMDCs

Bone marrow cells were collected from naive mice and grown for 8 days in RPMI supplemented with 10% FCS, 1% l-g lutamine, 1% sodium pyruvate, 0.1% 2-ME, 50 µg/mL streptomycin, 50 IU/mL penicillin, and 20 ng/mL recombinant murine GM-CSF (provided by Pr. Kris Thielemans, Medical School of the Vrije

Universiteit Brussel). At day 8, bone marrow-derived dendritic cells (BMDCs) were pulsed with 30 µg/mL KLH in the presence of 1 µg/mL LPS. At day 9, BMDCs were collected and injected in recipient mice (5 × 10^5 cells/f.p.).

Antibody Detection

Serum levels of NP-specific antibodies were determined by enzyme-linked immunosorbent assay (ELISA) according to standard procedures. Briefly, ELISA plates were coated with 2 µg/ml NP-BSA and incubated with serial dilutions of sera in duplicate wells. Bound antibodies were revealed using peroxidase-coupled anti-mouse isotype-specific rat monoclonal antibodies (Synabs sa, Louvain-la Neuve, Belgium) followed by the peroxidase substrate tetramethylbenzidine (Life Technologies). A solution of 2 N H$_2$SO$_4$ was used to quench the reaction, and optical densities were quantified at 450 nm. ODs were converted to units based on a standard curve made from a previously available immunized serum arbitrarily defined at 1,000 U ml^{-1}.

The relative affinities of NP-immune sera were calculated by comparing their binding to differently haptenised carrier proteins (heavily haptenised NP$_{18}$-BSA versus lightly haptenised NP$_2$-BSA; Biosearch Technologies, Inc.), as described (15). The same serial dilutions of each serum sample were allowed to bind on NP$_{18}$-BSA and NP$_2$-BSA. The relative affinities of the anti-NP serum antibodies are expressed as a ratio of the serum volumes required to give the 50% of maximum binding on NP$_{18}$-BSA divided by the volumes required for same binding on NP$_2$-BSA (serum relative affinity = vol$_{50\% binding}$ on NP$_{18}$-BSA/vol$_{50\% binding}$ on NP$_2$-BSA).

Flow Cytometry

Specific cell-surface staining was performed using a standard procedure with anti-CD4, anti-PD1, anti-IgD, anti-GL7 (eBioscience), and anti-CXCR5 mAbs (BD Biosciences).

Intracellular c-Maf, GATA-3, FoxP3, BCL6, Ki67 (Ab from BD Biosciences), and T-bet (eBioscience), staining was performed according to the manufacturer's protocol (FoxP3 staining set protocol, eBioscience). Cells were analyzed by flow cytometry with a FACS Canto II (BD Biosciences) and analyzed with the FlowJo Software. Live cells were analyzed within a FSC-A/FSC-H gate to exclude cell doublets and triplets.

Differentiation of Th Cells *In Vitro* and B-Cell Help Assay

CD62L^{hi}CD4$^+$ T cells and B cells were purified from naive animals by magnetic separation, as previously described (16, 17). The percentage of purified cell fractions in all experiments ranged between 90 and 98%, as estimated by flow cytometry (data not shown). Naive CD62L^{hi}CD4$^+$ T cells (5 × 10^5 cells/well in 24-well plates) were activated for 48–72 h with plastic-coated anti-CD3 mAb (5 µg/mL) and soluble anti-CD28 mAb (1 µg/mL). Th1 cells were differentiated by the addition of recombinant IL-12 (10 ng/mL) and anti-IL-4 mAb (10 µg/mL), and Th2 polarization was initiated by the addition of IL-4 (10 ng/mL) and anti-interferon-γ (IFN-γ) mAb (10 µg/mL). Th17 and Tfh cells were differentiated in media containing IL-6 (20 ng/mL) and supplemented (Th17) or not (Tfh) with TGF-β (3 ng/mL), in the presence of anti-IL-4 and anti-IFN-γ mAbs.

Serial dilutions of *in vitro*-derived Tfh cells were co-cultured for 7 days with syngeneic naïve B cells (5×10^4 cells/well) in the presence of anti-CD3 mAbs (500 ng/mL). T cells were irradiated (2,000 cGy) before the beginning of the coculture with B cells to prevent their outgrowth during the 7-day culture. IgG1 antibodies in the supernatants were determined by ELISA, using rat monoclonal anti-mouse isotype antibodies, as described (18) (Synabs, capture antibody loMG1.13; detection antibody loMK.1). Purified mouse IgG1 (BD Biosciences) was used as standard reference.

Bone Marrow Chimeras

Recipient CD3ε-KO mice were lethally irradiated with 2×600 rad and reconstituted with a 1:1 mixture of bone marrow from congenic wild type CD45.1[+] and c-Maf[KO-T] (CD45.2[+]) mice (2×10^6 cells of each). Chimeric mice were immunized with KLH/Alum 8 weeks after reconstitution.

Peyer's Patches Isolation

Peyer's patches (PPs) were cut from the small intestine and incubated for 20 min at 37°C in Hank's balanced salt solution (HBSS) containing 400 U/mL collagenase (Worthington, Lakewood, NJ, USA). PPs were next dissociated in HBSS/EDTA solution to generate a single-cell suspension for flow cytometry staining.

Real-time Quantitative RT-PCR

RNA was extracted using the TRIzol method (Invitrogen) and reverse transcribed with Superscript II reverse transcriptase (Invitrogen) according to the manufacturer's instructions. Quantitative real-time RT-PCR was performed using the SYBR Green Master mix kit (ThermoFisher). Primer sequences were as follow: RPL32 (F) ACATCGGTTATGGGAGCAAC; RPL32 (R) TCCAGCTCCTTGACATTGT; IL-4 (F) ATGCAC-GGAGATGGATGTG; IL-4 (R) AATATGCGAAGCACC-TTGGA; IL-17A (F) ATCCCTCAAAGCTCAGCGTGTC; IL-17A (R) GGGTCTTCATTGCGGTGGAGAG; IL-21 (F) GCCA-GATCGCCTCCTGATTA; IL-21(R) CATGCTCACAGTGCCCCTTT; IFNγ (F) TGCCAAGTTTGAGGTCAACA; IFNγ (R) GAATCAGCAGCGACTCCTTT.

Statistical Analysis

Differences between groups were analyzed with the Mann–Whitney test for two-tailed data. A p-value less than 0.05 was considered as significant.

RESULTS

c-Maf Is Expressed in Early Tfh Cells during *In Vivo* Vaccination Settings and Is Required for Tfh Cell Development

Keyhole limpet hemocyanin/Alum vaccination promotes the *in vivo* differentiation of cells expressing key Tfh-associated markers including PD1, CXCR5, and BCL6 by 72 h, gradually increasing till day 5 (**Figures 1A–D** and data not shown).

Kinetic analysis revealed that c-Maf was already expressed in the very few early Tfh found in the draining lymph nodes of day 3-vaccinated mice (**Figure 1E**). Although some c-Maf expression could be detected in the non-Tfh cell subset, higher levels of c-Maf were strongly associated with the Tfh cell compartment at any time points (**Figures 1E–J**). Moreover, c-Maf expression strongly correlated with BCL6 expression, therefore suggesting a time-coordinated expression pattern of these transcription factors throughout the Tfh cell differentiation process (**Figures 1E–L**).

To investigate the role of c-Maf in Tfh cell function and differentiation, we generated conditional KO mice in which *lox*P-flanked alleles of c-Maf (c-Maf[f/f]) are deleted by the Cre recombinase expressed from the CD4[+] T-cell-specific *Cd4* promoter (c-Maf[f/f]-CD4[CRE] mice, hereafter referred as c-Maf[KO-T] mice). These mice showed normal T cell development *in vivo* (data not shown) and naïve c-Maf-KO Th cells underwent optimal Th1, Th2, or Th17 polarization *in vitro* (**Figures 2A–C**). Upon culture in Tfh-promoting conditions, c-Maf-KO T cells displayed a twofold reduction in IL-21 production (**Figures 2D,E**), in agreement with a role for c-Maf in regulating IL-21 production (11). Finally, c-Maf-deficient Tfh cells showed reduced ability to deliver B cell help for IgG1 secretion *in vitro* (**Figure 2F**).

To assess the role of c-Maf in Tfh cell development, we immunized wild type and c-Maf[KO-T] mice with KLH/Alum or KLH/IFA/LPS. The loss of c-Maf strongly inhibited the differentiation of Tfh cells independently of the adjuvant formulation used (**Figures 3A–H**). The percentage and overall numbers of Tfh cells (assessed by either CXCR5[+] PD1[+], CD4[+] BCL6[+], or CXCR5[+]BCL6[+] T cells) were significantly decreased in c-Maf[KO-T] mice (**Figures 3B,C,E,F,H**).

Inhibition of Tfh cell development *in vivo* was already evidenced at early time points following immunization (days 2 and 4, Figure S1 in Supplementary Material) and did not result from a general impaired Th cell response as judged by the presence of proliferating T lymphocytes (as assessed by expression of the Ki67 marker, data not shown and **Figures 4D,E**) in the corresponding lymph nodes.

To test whether these observations were also applicable to non-vaccinal-driven Tfh responses, we analyzed the Tfh response that spontaneously occurs in the intestinal PP of mice harboring commensal microbiota. The proportion of Tfh cells among CD4[+] cell in PP was severely reduced in c-Maf[KO-T] mice (**Figures 3I–K**), suggesting that c-Maf expression in T cells was also essential for normal homeostatic Tfh cell responses.

c-Maf Promotes Tfh Cell Differentiation in a T-Cell Intrinsic Manner

To address the potential role of c-Maf in the regulation of Th cell responses *in vivo*, we immunized C57BL/6 mice with BMDCs loaded with KLH in the presence of LPS. We previously reported that this vaccination protocol leads to the coordinated upregulation of Tfh, Th1, Th2, and Th17 responses (19, 20). Analysis of the CD4[+] T cell response further confirmed the defect in Tfh cell differentiation despite normal accumulation of CD4[+]Ki67[+] cells in c-Maf[KO-T] mice (**Figures 4A–F,J**). Notably, CD4[+]Ki67[+] cells recovered from c-Maf[KO-T] immunized mice expressed higher levels

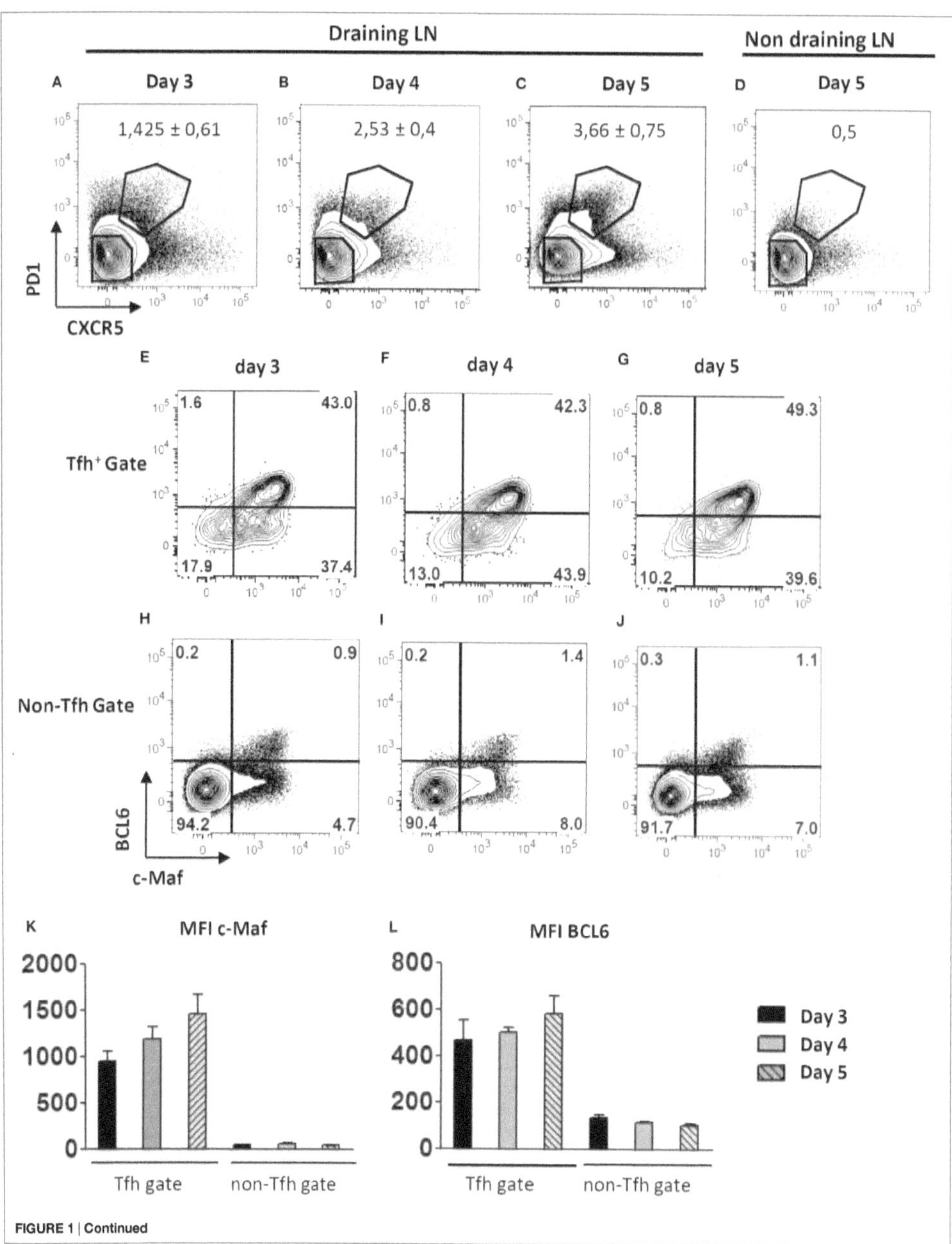

FIGURE 1 | Continued

FIGURE 1 | Continued
Early expression of c-Maf during the course of follicular helper T cells (Tfh) cell differentiation. WT mice were immunized by footpad injection of keyhole limpet hemocyanin in Alum. Draining lymph nodes were recovered on days 3–5 and analyzed for Tfh cell marker expression. Contour plots in **(A–D)** illustrate CXCR5 and PD1 staining profiles of CD4+ cells while contour plots in **(E–J)** represent BCL6 and c-Maf staining profiles of Tfh cells (CXCR5+ PD1+) and non-Tfh cells (CXCR5− PD1−) gated as indicated in **(A–D)**; **(K,L)** histograms of c-Maf or BCL6 expression intensity (expressed as mfi, median fluorescence intensity) among Tfh and non-Tfh cell subsets. Results are representative of two independent experiments; histograms and numbers in **(A–D)** represent the mean ± SD of three individual mice.

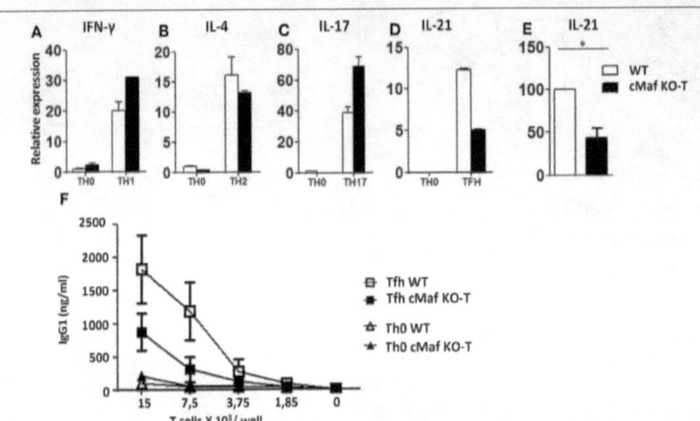

FIGURE 2 | c-Maf up-regulates IL-21 secretion and B-cell help capacity by follicular helper T cells (Tfh)-like cells *in vitro*. (A–E) Naïve T cells were purified from the spleen of C57BL/6 mice and stimulated for 2 days with anti-CD3/CD28 mAbs under Th0, Th1, Th2, Th17, or Tfh-like conditions. Gene expression relative to RPL32 mRNA of **(A)** interferon-γ, **(B)** IL-4, **(C)** IL-17, and **(D)** IL-21 was assessed by qRT-PCR. WT Th0 sample was set to 1. **(E)** Compilation of four independent experiments as in panel **(D)**: IL-21 expression in WT Tfh cells was set as 100. **(F)** Serial dilutions of Th0 and Tfh cells were incubated with purified B cells (5 × 10⁵ cells/well) and anti-CD3 mAbs (500 ng/ml). Culture supernatants were tested on day 7 for IgG1 content. Results are expressed as mean ± SD of duplicates **(A–D)** or four independent cultures **(F)** and are representative of at least two independent experiments. Differences between groups were analyzed with the Mann–Whitney test for two-tailed data. *$p < 0.05$.

of T-bet or GATA-3, the major transcription factors associated with the Th1- or Th2-cell differentiation program, respectively (**Figures 4H,I,L–O**). As expected from the literature (21), c-Maf[KO-T] mice failed to express optimal levels of RORγt following immunization (**Figures 4G,K**).

Overall, these observations point to a role for c-Maf in promoting Tfh and Th17 cell differentiation, while restraining the development of Th1 and Th2 lymphocytes upon antigen-stimulation *in vivo*.

We next generated chimeras by reconstituting irradiated CD3⁻/⁻ recipient mice with a 1:1 mixture of congenitally marked bone marrow cells from wild type (CD45.1⁺) and c-Maf[KO-T] (CD45.2⁺) donor mice. Despite an equivalent proliferative response upon immunization (**Figures 5A,B,E**), cMaf-deficient CD4⁺ T cells displayed a reduced potential to acquire Tfh-features, both in the periphery (lymph nodes draining the immunization site, **Figures 5C,D**) and in the PP (**Figure 5F**), pointing to a cell-autonomous defect in the Tfh differentiation of c-Maf-deficient CD4⁺ T cells.

c-Maf Expression in T Cells Is Required for Optimal Humoral Responses

We next tested whether c-Maf deficiency in the T cell compartment would affect the development of humoral responses. The loss of c-Maf led to a strongly reduced GC B cell response in the draining lymph nodes of immunized mice (**Figure 6A**). Accordingly, NP-KLH/Alum immunized c-Maf[KO-T] mice produced lower anti-NP IgG1 antibodies titers at all time points (primary to late secondary response) considered, whereas anti-NP IgM levels remained non-significantly affected (**Figures 6B–D,F**). In particular, mice lacking c-Maf expression in T cells failed to secrete detectable levels of high-affinity NP-specific IgG1 antibodies (**Figures 6E,G,H**), in keeping with the concept that c-Maf-signaling in T cells promotes Tfh cell development, thereby controlling both isotype switch and affinity maturation processes.

DISCUSSION

While expression of cMaf has long been considered as a hallmark of Tfh cells, its precise role during Tfh cell development was unknown at the beginning of the present study. We demonstrate herein that c-Maf plays a major, non-redundant role in the differentiation of Tfh cells *in vivo*. c-Maf is expressed early during Tfh cell induction and mice harboring a T cell-specific conditional ablation of c-Maf are strongly impaired in their Tfh response to both antigen/adjuvant vaccinations and commensal intestinal bacteria.

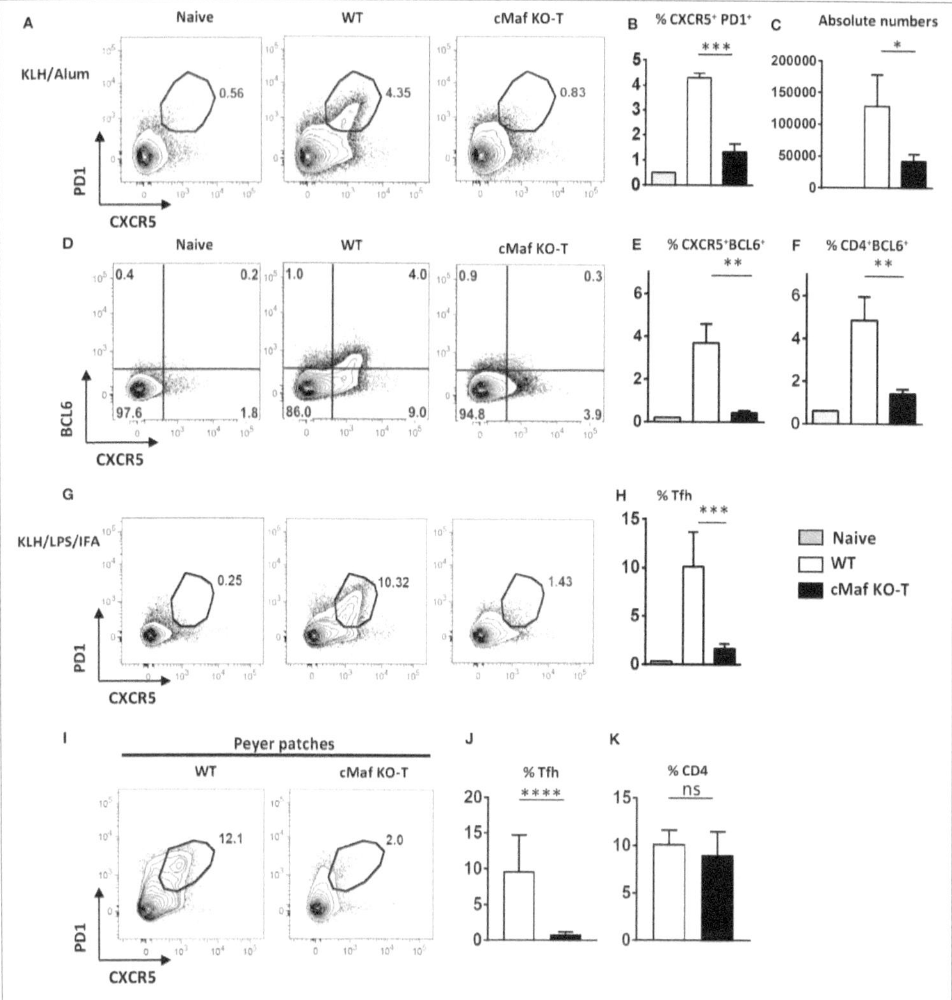

FIGURE 3 | c-Maf is required for follicular helper T cells (Tfh) cell development. (A–H) c-Maf$^{fl/fl}$ and c-Maf^{KO-T} mice were immunized with keyhole limpet hemocyanin (KLH) in Alum **(A–F)** or KLH + LPS in IFA **(G,H)**. Draining lymph nodes were analyzed on day 7 for Tfh cells (defined as CXCR5+ PD1+, CD4+ BCL6+, or CXCR5+ BCL6+) among viable CD4+ cells. **(A,D,G)** show representative contour plots; histograms **(B,C,E,F,H)** represent the percentages, and absolute numbers of Tfh cells, as indicated. **(I–K)** Tfh (CXCR5+ PD1+) and total CD4 T cell expression in Peyer's patches of wild type and c-Maf^{KO-T} mice. Results represent the mean ± SD of four to five individual mice and are representative of at least three independent experiments **(B,C,E,F,H)** or are pooled from four independent experiments [n = 11–13 **(J,K)**]. Differences between groups were analyzed with the Mann–Whitney test for two-tailed data. $^{*}p < 0.05$; $^{**}p < 0.01$; $^{***}p < 0.001$; $^{****}p < 0.0001$.

By using retroviral ectopic expression of c-Maf or BCL6 in *in vitro*-derived human Tfh cells, Kroenke et al. first reported that c-Maf and BCL6 regulated distinct features of Tfh cell function, with BCL6 being critical for T cell positioning within the follicle and directing T–B interactions (through upregulation of CXCR5, CXCR4, CCR7, SAP, PD1, CD40L, ICOS, CXCL13) and c-Maf promoting and sustaining IL-21 secretion (10). The cooperation of c-Maf and BCL6/Ascl2 to achieve complete Tfh cell fate through orchestration of distinct sets of genes was further supported by studies indicating that c-Maf-bound genes hardly

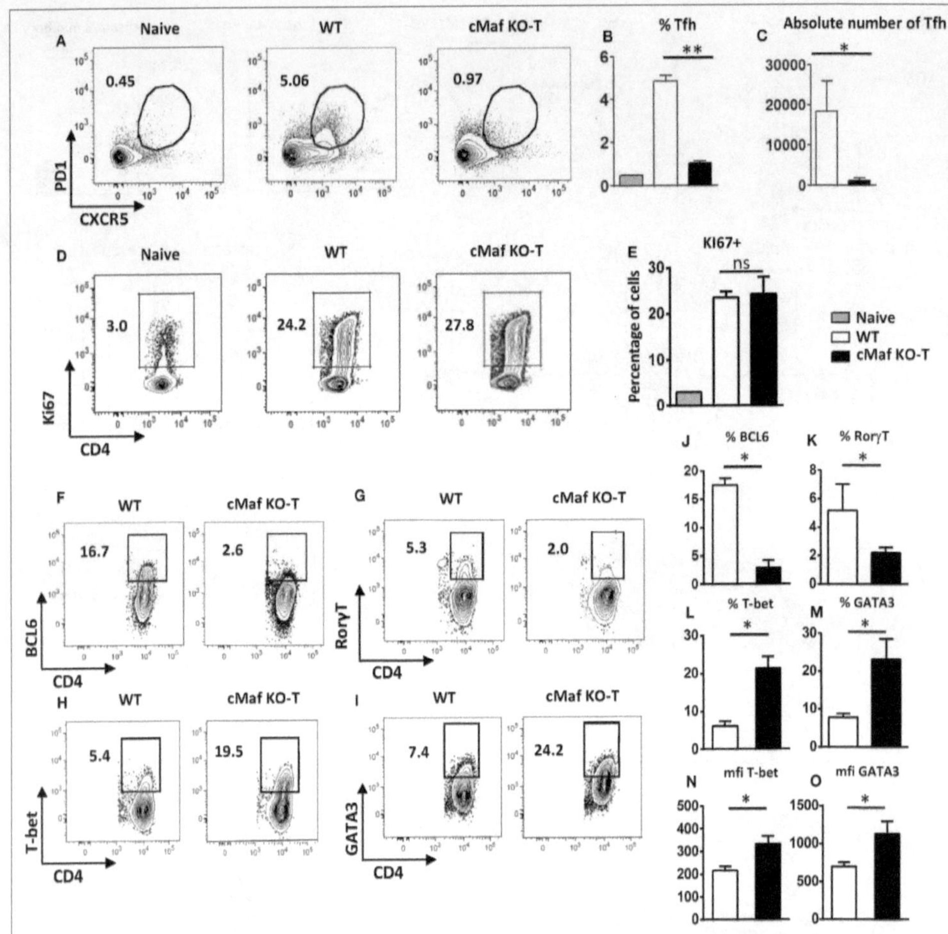

FIGURE 4 | c-Maf promotes follicular helper T cells (Tfh) cell differentiation and counteracts Th1 and Th2 cell development. WT mice were immunized by footpad injection of 5 × 10⁶ LPS-stimulated, KLH-pulsed bone marrow-derived dendritic cells derived from C57BL/6 mice. Draining LN cells were recovered on day 7 and analyzed for Tfh cell expression **(A–C)** and Ki67 proliferation marker expression **(D,E)** among viable CD4⁺ T cells or BCL6, RORγt, T-bet, and GATA-3 expression among CD4⁺Ki67⁺ cell gate **(F–I)**. Histograms represent the percentage or absolute numbers of Tfh among viable CD4⁺ T cell gate **(B,C,E)**, percentages of cells among CD4⁺Ki67⁺ cell gate **(J–M)**, or transcription factor expression intensity [mfi **(N,O)**]. Differences between groups were analyzed with the Mann–Whitney test for two-tailed data. *$p < 0.05$; **$p < 0.01$; ***$p < 0.001$.

correlated with genes bound by BCL6 or Ascl2 (8, 9). Collectively, and in agreement with the well-known role of c-Maf in transactivating the IL-4 and IL-21 promoters (11, 12), these data suggested that c-Maf regulates Tfh cell function by fine-tuning cytokine production in Ascl2/BCL6-driven differentiated Tfh cells (1). Our data challenge this simple view of cMaf sustaining the Tfh program through cytokine, and in particular IL-21, production only. Indeed, c-Maf deficiency in the T cell compartment led to a virtually complete loss of Tfh cells, while lack of IL-21–IL-21R signaling appeared as largely compatible with Tfh cell development (22–24). Similarly, reduction in Tfh cell numbers was more severe in c-Maf^KO-T than in mice lacking STAT3 (the major IL-6/IL-21 signaling pathway in T cells) (20, 22, 25), thus suggesting that the defect observed in c-Maf^KO-T mice did not simply reflect a consequence of reduced IL-21 signaling in these cells.

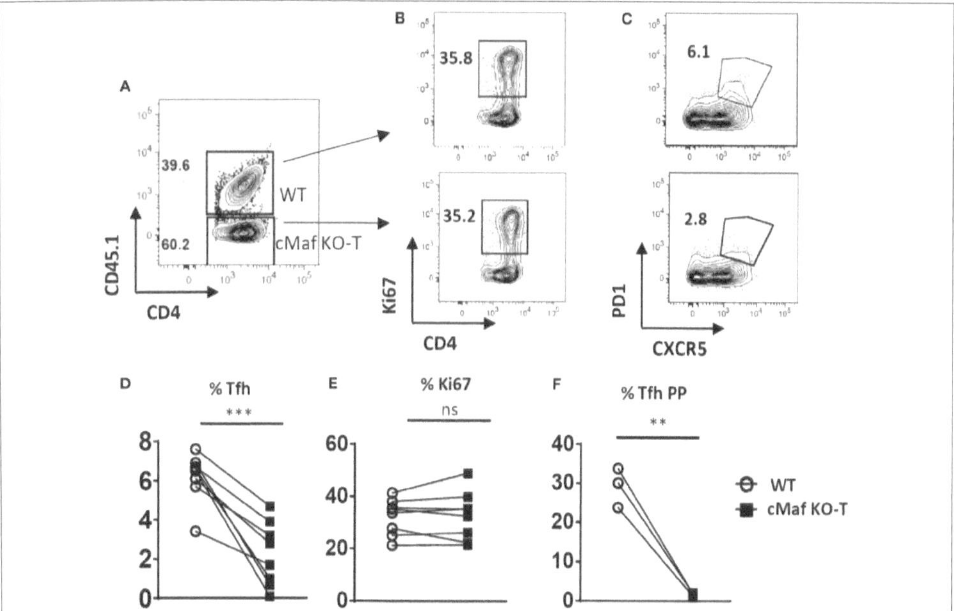

FIGURE 5 | c-Maf induces follicular helper T cells (Tfh) cell differentiation in a T-cell intrinsic manner. Flow cytometry of CD4+ T cells in the spleen of chimeras generated with a mixture of WT (CD45.1) and c-Maf[KO-T] (CD45.2) bone marrow cells, assessed 7 days after immunization with keyhole limpet hemocyanin/ Alum in f.p. **(A)** gating strategy to identify WT and c-Maf-KO CD4+ T cells in the chimeric mice; **(B,C)** representative contour plots of activation and Tfh markers in both WT and c-Maf-KO CD4+ T cell subsets; **(D,E)** summary of Tfh and activated cell frequency from **(B,C)**; **(F)** summary of Tfh cell frequency in Peyer's patches of chimeras. Each symbol represents individual mice. Results represent the mean ± SD of eight individual mice and are pooled from two independent experiments **(D,E)** or are representative of two independent experiments **(F)**. Paired Mann–Whitney test: **$p < 0.01$; ***$p < 0.001$.

Using a fetal liver transplant experiment, Bauquet et al. reported that loss of c-Maf resulted in markedly reduced numbers of Th17 and Tfh cells (26), suggesting that c-Maf might also regulate Tfh cell differentiation. However, the entire hematopoietic compartment (including T, B, and APCs) was defective for c-Maf in these chimeric mice, still leaving open the question of a direct intrinsic role of c-Maf in the Tfh cell differentiation program. In keeping with these conclusions, we provide evidence herein suggesting that c-Maf controls Tfh cell fate in a cell-autonomous fashion. Indeed, wild type Tfh cells failed to rescue Tfh differentiation of c-Maf-deficient T cells in chimeric mice, suggesting that soluble factors provided by wild type T cells cannot compensate for c-Maf loss during Tfh cell development.

As mentioned earlier, a first wave of pre-Tfh cell induction precedes the full Tfh cell differentiation process (characterized by stabilization of high levels of BCL6 expression) (5–7). Our *in vivo* studies reveal that c-Maf and BCL6 are co-expressed during this early step of Tfh cell development. Although previous studies and our own unpublished data have shown that over-expression of c-Maf does not directly induce BCL6 expression in Th cells (8, 10), we observe a severe defect in BCL6 expression in CD4+ Th cells,

1 week after immunization, in the absence of c-Maf. Collectively, these data thus suggest that promotion of the early wave of Tfh cell differentiation by c-Maf may facilitate further stabilization of BCL6 expression in fully committed Tfh cells.

In vitro studies confirmed that c-Maf expression was required for adequate IL-21 production (11). As previously documented, activation of naïve Th cells in the presence of IL-6 led to the differentiation of IL-21-producing Tfh-like cells able to induce B cell growth and antibody secretion (17, 18). Using the same *in vitro* assay, we observed a twofold reduction in IL-21 secretion by c-Maf-deficient Tfh-like cells, concomitantly with a reduced ability to deliver B cell help for antibody production in T–B co-culture assays. Thus, although failure to secrete adequate IL-21 levels may explain the reduced B cell help properties of c-Maf-deficient T cells *in vitro*, the strict requirement for c-Maf expression to achieve B cell help *in vivo* probably reflects additional roles for this transcription factor in driving Tfh development in a physiological setting.

The negative influence of c-Maf on Th2 responses may appear at odds with previous publications that led several groups to consider c-Maf as a prototypical Th2-transcription factor (27, 28).

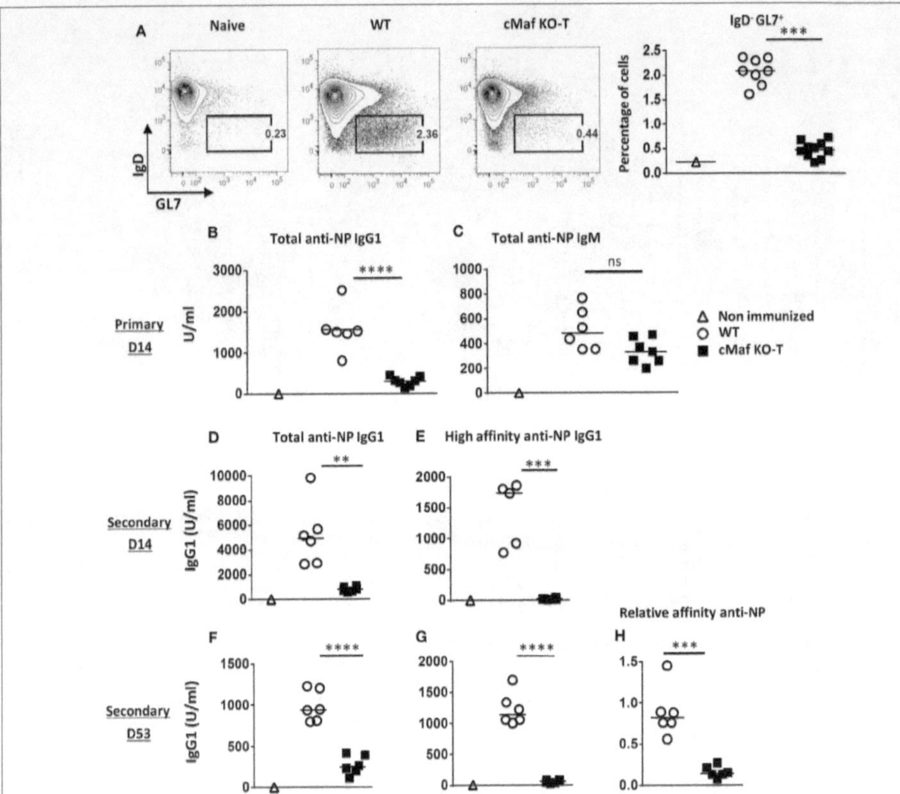

FIGURE 6 | c-Maf is required for high-affinity antibody response. (A) Control and c-Maf[KO-T] mice were immunized with keyhole limpet hemocyanin (KLH)/Alum in fp and draining lymph node were analyzed on day 7. Contour plots and histograms represent B_{GC} cells (defined as IgD[low]-GL7+, among CD19+ B cell gate). **(B–H)** Mice were immunized ip with NP-KLH in Alum. Sera were collected at the indicated timings and tested for total [fixation on NP_{24}-BSA-coated enzyme-linked immunosorbent assay (ELISA) plates] anti-NP IgG1 **(B,D,F)** and IgM **(C)**. High-affinity IgG1 were detected on NP_4-BSA-coated ELISA plates **(E,G)**. Relative affinity is expressed as the ratio of serum volume corresponding to 50% binding on NP_{24}-BSA and NP_4-BSA **(H)**. Each symbol represents individual mice. Results are pooled from two independent experiments **(A)** or are representative of three independent experiments **(B–H)**. Difference between groups was analyzed with the Mann–Whitney test for two-tailed data. **$p < 0.01$; ***$p < 0.001$; ****$p < 0.0001$.

Notably, most of these studies were based on *in vitro* assays suggesting a positive role for c-Maf in regulating IL-4 promoter activity. However, Sato et al. later reported that c-Maf-transgenic memory T helper cells did not produce enhanced IL-4 levels when compared to WT cells, and that the expression of *Il4* was not reduced in immunodeficient mice reconstituted with c-Maf$^{-/-}$ fetal liver transplants (29). In keeping with this last study, we observed an enhanced, rather than reduced, production of IL-4 by c-Maf-deficient T cells induced *in vivo* in response to antigen-pulsed DC vaccination. Similarly, c-Maf-deficient T cells displayed a marked Th1-like phenotype, further suggesting a possible role for c-Maf in controlling non-Tfh responses. These

observations may suggest that similarly to BCL6, c-Maf may promote Tfh differentiation by limiting the acquisition of Th1/Th2 phenotypic traits. The identification of the precise mechanism whereby c-Maf may limit Th2 responses was beyond the scope of this study but might be related to limitation of GATA-3 expression in Th2 cells in our *in vivo* setting. Noteworthy, both GATA-3 and cMaf are involved in vertebrate lens development (30). Of interest, GATA-3 expression is strongly upregulated in c-Maf-deficient lens, a finding compatible with a negative role for c-Maf in regulating GATA-3 expression (31).

In conclusion, the fact that Tfh cell differentiation strongly relies on ICOSL/ICOS and IL-6/STAT3 signaling—two pathways known to induce expression of c-Maf (18, 32–34)—strongly supports our findings indicating an important role for c-Maf in driving Tfh cell differentiation. The present study thus adds c-Maf to a lengthy list of transcription factors (including BCL6, Ascl2, TCF-1, Batf, STAT3, NFAT, and IRF4) that have been shown to play a major, non-redundant role in Tfh cell development (1). The identification of the relative roles of these transcription factors in initiating and maintaining Tfh cell represents therefore an important challenge requiring further studies.

ETHICS STATEMENT

The experiments were carried out in strict accordance with the relevant laws of the country and with institutional guidelines. We received specific approval for this study from the Université Libre de Bruxelles Institutional Animal Care and Use Committee (protocol numbers CEBEA-5, 31, and 92).

AUTHOR CONTRIBUTIONS

FA conceived and supervised the research program and experiments, acquired and analyzed data, and wrote the manuscript. SD, MA, MH, and HH performed the experiments and contributed to the analysis of the data. OL conceived the research program, provided financial support, and revised the manuscript.

ACKNOWLEDGMENTS

We thank Caroline Abdelaziz and Véronique Dissy for animal care and for technical support. We also thank Dr. Carmen Birchmeier (Max Delbrück Center for Molecular Medicine, Berlin, Germany) for generously providing the recombinant c-Maf-flox mice. FA specially thanks Muriel Moser for her great support all along this work and for critical reviewing of the manuscript.

FUNDING

This work was supported by The Belgian Program in Interuniversity Poles of Attraction Initiated by the Belgian State, Prime Minister's office, Science Policy Programming, by a Research Concerted Action of the Communauté Française de Belgique, grant from the Fonds Jean Brachet and research credit from the National Fund for Scientific Research, FNRS, Belgium. FA is a Research Associate at the FNRS. MA and MH were supported by a Belgian FRIA fellowship.

SUPPLEMENTARY MATERIAL

The Supplementary Material for this article can be found online at http://journal.frontiersin.org/article/10.3389/fimmu.2017.00480/full#supplementary-material.

REFERENCES

1. Vinuesa CG, Linterman MA, Yu D, MacLennan I. Follicular helper T cells. *Annu Rev Immunol* (2016) 34:335–68. doi:10.1146/annurev-immunol-041015-055605

2. Johnston RJ, Poholek AC, DiToro D, Yusuf I, Eto D, Barnett B, et al. Bcl6 and Blimp-1 are reciprocal and antagonistic regulators of T follicular helper cell differentiation. *Science* (2009) 325(5943):1006–10. doi:10.1126/science.1175870

3. Nurieva RI, Chung Y, Martinez GJ, Yang XO, Tanaka S, Matskevitch TD, et al. Bcl6 mediates the development of T follicular helper cells. *Science* (2009) 325(5943):1001–5. doi:10.1126/science.1176676

4. Linterman MA, Beaton L, Yu D, Ramiscal RR, Srivastava M, Hogan JJ, et al. IL-21 acts directly on B cells to regulate Bcl-6 expression and germinal center responses. *J Exp Med* (2010) 207(2):353–63. doi:10.1084/jem.20091738

5. Choi YS, Kageyama R, Eto D, Escobar TC, Johnston RJ, Monticelli L, et al. ICOS receptor instructs T follicular helper cell versus effector cell differentiation via induction of the transcriptional repressor Bcl6. *Immunity* (2011) 34(6):932–46. doi:10.1016/j.immuni.2011.03.023

6. Liu D, Xu H, Shih C, Wan Z, Ma X, Ma W, et al. T-B-cell entanglement and ICOSL-driven feed-forward regulation of germinal centre reaction. *Nature* (2015) 517(7533):214–8. doi:10.1038/nature13803

7. Baumjohann D, Okada T, Ansel KM. Cutting edge: distinct waves of Bcl6 expression during T follicular helper cell development. *J Immunol* (2011) 187(5):2089–92. doi:10.4049/jimmunol.1101393

8. Liu X, Chen X, Zhong B, Wang A, Wang X, Chu F, et al. Transcription factor achaete-scute homologue 2 initiates follicular T-helper-cell development. *Nature* (2014) 507(7493):513–8. doi:10.1038/nature12910

9. Liu X, Lu H, Chen T, Nallaparaju KC, Yan X, Tanaka S, et al. Genome-wide analysis identifies Bcl6-controlled regulatory networks during T follicular helper cell differentiation. *Cell Rep* (2016) 14(7):1735–47. doi:10.1016/j.celrep.2016.01.038

10. Kroenke MA, Eto D, Locci M, Cho M, Davidson T, Haddad EK, et al. Bcl6 and Maf cooperate to instruct human follicular helper CD4 T cell differentiation. *J Immunol* (2012) 188(8):3734–44. doi:10.4049/jimmunol.1103246

11. Hiramatsu Y, Suto A, Kashiwakuma D, Kanari H, Kagami S, Ikeda K, et al. c-Maf activates the promoter and enhancer of the IL-21 gene, and TGF-beta inhibits c-Maf-induced IL-21 production in CD4+ T cells. *J Leukoc Biol* (2010) 87(4):703–12. doi:10.1189/jlb.0909639

12. Sahoo A, Alekseev A, Tanaka K, Obertas L, Lerman B, Haymaker C, et al. Batf is important for IL-4 expression in T follicular helper cells. *Nat Commun* (2015) 6:7997. doi:10.1038/ncomms8997

13. Wende H, Lechner SG, Cheret C, Bourane S, Kolanczyk ME, Pattyn A, et al. The transcription factor c-Maf controls touch receptor development and function. *Science* (2012) 335(6074):1373–6. doi:10.1126/science.1214314

14. Lee PP, Fitzpatrick DR, Beard C, Jessup HK, Lehar S, Makar KW, et al. A critical role for Dnmt1 and DNA methylation in T cell development, function, and survival. *Immunity* (2001) 15(5):763–74. doi:10.1016/S1074-7613(01)00227-8

15. Eddahri F, Oldenhove G, Denanglaire S, Urbain J, Leo O, Andris F. CD4+ CD25+ regulatory T cells control the magnitude of T-dependent humoral immune responses to exogenous antigens. *Eur J Immunol* (2006) 36(4):855–63. doi:10.1002/eji.200535500

16. Andris F, Van Mechelen M, De Mattia F, Baus E, Urbain J, Leo O. Induction of T cell unresponsiveness by anti-CD3 antibodies occurs independently of co-stimulatory functions. *Eur J Immunol* (1996) 26(5):1187–95. doi:10.1002/eji.1830260534

17. Mari N, Hercor M, Denanglaire S, Leo O, Andris F. The capacity of Th2 lymphocytes to deliver B-cell help requires expression of the transcription factor Stat3. *Eur J Immunol* (2013) 43(6):1489–98. doi:10.1002/eji.201242938

18. Eddahri F, Denanglaire S, Bureau F, Spolski R, Leonard WJ, Leo O, et al. Interleukin-6/Stat3 signaling regulates the ability of naive T cells to acquire B-cell help capacities. *Blood* (2009) 113(11):2426–33. doi:10.1182/blood-2008-04-154682

19. Mayer A, Debuisson D, Denanglaire S, Eddahri F, Fievez L, Hercor M, et al. Antigen presenting cell-derived IL-6 restricts Th2-cell differentiation. *Eur J Immunol* (2014) 44(11):3252–62. doi:10.1002/eji.201444646

20. Hercor M, Anciaux M, Denanglaire S, Debuisson D, Leo O, Andris F. Antigen-presenting cell-derived IL-6 restricts the expression of GATA3 and IL-4 by follicular helper T cells. *J Leukoc Biol* (2017) 101(1):5–14. doi:10.1189/jlb.1HI1115-511R

21. Tanaka S, Suto A, Iwamoto T, Kashiwakuma D, Kagami S, Suzuki K, et al. Sox5 and c-Maf cooperatively induce Th17 cell differentiation via RORγt induction as downstream targets of Stat3. *J Exp Med* (2014) 211(9):1857–74. doi:10.1084/jem.20130791

22. Zotos D, Coquet JM, Zhang Y, Light A, D'Costa K, Kallies A, et al. IL-21 regulates germinal center B cell differentiation and proliferation through a B cell-intrinsic mechanism. *J Exp Med* (2010) 207(2):365–78. doi:10.1084/jem.20091777

23. Eto D, Lao C, DiToro D, Barnett B, Escobar TC, Kageyama R, et al. IL-21 and IL-6 are critical for different aspects of B cell immunity and redundantly induce optimal follicular helper CD4 T cell (Tfh) differentiation. *PLoS One* (2011) 6(3):e17739. doi:10.1371/journal.pone.0017739

24. Karnowski A, Chevrier S, Belz GT, Mount A, Emslie D, D'Costa K, et al. B and T cells collaborate in antiviral responses via IL-6, IL-21, and transcriptional activator and coactivator, Oct2 and OBF-1. *J Exp Med* (2012) 209(11):2049–64. doi:10.1084/jem.20111504

25. Wu H, Xu L-L, Teuscher P, Liu H, Kaplan MH, Dent AL. An inhibitory role for the transcription factor Stat3 in controlling IL-4 and Bcl6 expression in follicular helper T cells. *J Immunol* (2015) 195(5):2080–9. doi:10.4049/jimmunol.1500335

26. Bauquet AT, Jin H, Paterson AM, Mitsdoerffer M, Ho IC, Sharpe AH, et al. The costimulatory molecule ICOS regulates the expression of c-Maf and IL-21 in the development of follicular T helper cells and Th-17 cells. *Nat Immunol* (2009) 10(2):167–75. doi:10.1038/ni.1690

27. Ho IC, Hodge MR, Rooney JW, Glimcher LH. The proto-oncogene c-Maf is responsible for tissue-specific expression of interleukin-4. *Cell* (1996) 85(7):973–83. doi:10.1016/S0092-8674(00)81299-4

28. Kim JI, Ho I-C, Grusby MJ, Glimcher LH. The transcription factor c-Maf controls the production of interleukin-4 but not other Th2 cytokines. *Immunity* (1999) 10(6):745–51. doi:10.1016/S1074-7613(00)80073-4

29. Sato K, Miyoshi F, Yokota K, Araki Y, Asanuma Y, Akiyama Y, et al. Marked induction of c-Maf protein during Th17 cell differentiation and its implication in memory Th cell development. *J Biol Chem* (2011) 286(17):14963–71. doi:10.1074/jbc.M111.218867

30. Cvekl A, Ashery-Padan R. The cellular and molecular mechanisms of vertebrate lens development. *Development* (2014) 141(23):4432–47. doi:10.1242/dev.107953

31. Maeda A, Moriguchi T, Hamada M, Kusakabe M, Fujioka Y, Nakano T, et al. Transcription factor GATA-3 is essential for lens development. *Dev Dyn* (2009) 238(9):2280–91. doi:10.1002/dvdy.22035

32. Nurieva RI, Mai XM, Forbush K, Bevan MJ, Dong C. B7h is required for T cell activation, differentiation, and effector function. *Proc Natl Acad Sci U S A* (2003) 100(24):14163–8. doi:10.1073/pnas.2335041100

33. Yang Y, Ochando J, Yopp A, Bromberg JS, Ding Y. IL-6 plays a unique role in initiating c-Maf expression during early stage of CD4 T cell activation. *J Immunol* (2005) 174(5):2720–9. doi:10.4049/jimmunol.174.5.2720

34. Mak TW, Shahinian A, Yoshinaga SK, Wakeham A, Boucher LM, Pintilie M, et al. Costimulation through the inducible costimulator ligand is essential for both T helper and B cell functions in T cell-dependent B cell responses. *Nat Immunol* (2003) 4(8):765–72. doi:10.1038/ni947

Conflict of Interest Statement: The authors declare no conflict of interest regarding the publication of this paper. The funders had no role in study design, data collection and analysis, decision to publish, or preparation of the manuscript.

Copyright © 2017 Andris, Denanglaire, Anciaux, Hercor, Hussein and Leo. This is an open-access article distributed under the terms of the Creative Commons Attribution License (CC BY). The use, distribution or reproduction in other forums is permitted, provided the original author(s) or licensor are credited and that the original publication in this journal is cited, in accordance with accepted academic practice. No use, distribution or reproduction is permitted which does not comply with these terms.

www.ingramcontent.com/pod-product-compliance
Lightning Source LLC
LaVergne TN
LVHW042113190726
843493LV00006B/1473